T0092926

Data Plane Development KIT (DPDK)

Data Plane Development KIT (DPDK)

A Software Optimization Guide to the User Space-based Network Applications

Edited by
Heqing Zhu

CRC Press
Taylor & Francis Group
Boca Raton London New York

CRC Press is an imprint of the
Taylor & Francis Group, an **informa** business

First edition published 2021
by CRC Press
6000 Broken Sound Parkway NW, Suite 300, Boca Raton, FL 33487-2742

and by CRC Press
2 Park Square, Milton Park, Abingdon, Oxon, OX14 4RN

© 2021 Taylor & Francis Group, LLC

CRC Press is an imprint of Taylor & Francis Group, LLC

ISBN: 9780367520175 (hbk)
ISBN: 9780367373955 (pbk)
ISBN: 9780429353512 (ebk)

Typeset in Times
by codeMantra

DPDK uses the FreeBSD license for the most software code, which are running in the user mode. A small amount of code will reside in the kernel mode such as VFIO, KNI, which are released under GPL. BSD license model gives the developers and consumers more flexibility for the commercial use. If a developer wants to contribute any new software code, the license model need be followed. If we go to the DPDK website (www.dpdk.org), we can download the source code for immediate use.

DPDK open source development is still very active today, rapidly evolving with new features and large source code contribution, the open source package is released every 3 months, this release cadence is decided since 2016.

Contents

SECTION 1   DPDK Foundation

SECTION 2   I/O Virtualization

SECTION 3   DPDK Applications

Preface

DPDK (Data Plane Development Kit) started as a small software project at Intel® about a decade ago. By 2019, it has evolved into a leading open-source project under the governance of Linux Foundation. DPDK is known as a kernel bypass networking technology, and it has gained the huge adoption by cloud, telecom, and enterprise networking and security systems. Many technical articles and online documentation shared the DPDK technology and application; many of the contents are available on the Internet. They are very informative and useful, but isolated in many different places. As of 2019, the general-purpose processors are widely used for cloud and telecom network infrastructure systems. DPDK played an important role to deliver the high-speed I/O, and it is widely used in software-defined data centers (Open vSwitch, VMware NSX, Red Hat virtualization, load balancers, etc.) and telecom networking appliances such as Virtual EPC, Virtual Router, and NGFW (next-generation firewall).

In April 2015, the first DPDK China summit was held in Beijing. Network developers and system designers from China Mobile, China Telecom, Alibaba, IBM, Intel®, Huawei, and ZTE presented a a variety of topics around DPDK technologies and their use cases. The passionate audience inspired us to write this book about DPDK, and the first edition of Chinese book was published in 2016. In the past 3 years, there were continuous improvements in DPDK community. We incorporate some changes into this English book.

The DPDK community spent lots of effort on the documentation improvement; the significant progress has been made. This book is intended to make DPDK easy to use, and to cover the design principles and software libraries and software optimization approach. We have to admit we have the limited knowledge and language skills, which may leave this book imperfect. This book is a joint work from many talented engineers from Intel®, Alibaba, and Huawei; they are DPDK developers and users. For network engineers and college students, if you are working to build a networking/security system or service, which has not used DPDK yet, this book may help you.

The "network function virtualization (NFV)" transformation has inspired DPDK as a key data plane technology, a crucial ingredient to build the 5G and edge systems (uCPE/SD-WAN, MEC). DPDK is used as a foundational I/O software to deliver the wireless and fixed network services.

SILICON

When I joined Intel® in 2005, a big challenge is how to ensure software to gain the benefits using multicore CPU. Most legacy software was not designed with parallelization in mind, not able to gain performance benefits from the multicore processor. Prior to 2005, CPU design focused on increasing the running frequency; the software gained performance improvement without any changes; the increased CPU frequency helps the software to achieve a higher performance. As industry enters

TABLE 1

The Intel® Xeon Processor in Years

CPU Code Name	CPU Process (nm)	Max # Cores per CPU	Release Time	Hyper-Threading	Total # of Cores on 2-CPU Server
WoodCrest	65	2	2006	No	4
Nehalem-EP	45	4	2009	Yes	16
Westmere-EP	32	6	2010	Yes	24
Sandy Bridge-EP	32	8	2012	Yes	32
Ivy Bridge-EP	22	12	2013	Yes	48
Haswell-EP	22	18	2014	Yes	72
Skylake-EP	14	28	2017	Yes	112
Cascade Lake-EP	14	56	2019	Yes	224

the multicore journey (almost 15 years up to now), the CPU innovation focuses on delivering more cores at each new product release cycle. Table 1 summarizes the core count increase history on Intel® Xeon processors.

The server platform evolution is largely driven by Intel® Xeon processor; the most commonly used server platform is based on the dual-socket design; the number of logical computing cores increased about 56 times within 13 years. It is a similar aggressive technology development from all other CPU suppliers. In addition to the rapid CPU development, the other silicon components in the server system, such as memory and NIC (network interface card), have gone through a similar and significant capability boost.

Ethernet technology has evolved from the early interface speed at 10/100 Mbps to gigabit Ethernet (1 Gbps). Today, 10 Gbps/25 Gbps NIC is adopted by the new data center servers. In the networking production systems, 100 Gbps ultra-high-speed network interface is also being used, but CPU frequency has remained at the same level as 10 years ago. Table 2 summarizes Ethernet development history from https://ethernetalliance.org/.

Supporting the high-speed NIC, substantial software innovations should be made; it is important to have the software innovation with the high parallelism. DPDK is born in the time of change; one of the designed goals is to take advantage of the multicore architecture to enable high-speed NIC.

DPDK is an open-source project with broad industry support; many organizations recognize the power of open source and start to participate and contribute; there is a very clear progress in cloud and telecom/security solution. DPDK is born at Intel® first, but it gained the broad contribution due to the open source model.

From a cost perspective, the server platform cost is reduced significantly, and a dual-socket server today is just about the similar cost to a high-end laptop 10 years ago. The computing capability exceeds what a supercomputer can do at that time. Such rapid technology development makes the server platform as the preferable

TABLE 2
Ethernet Port Speed Evolution

Ethernet Port Speed	Year
10 Mb/s	~1980
100 Mb/s	~1995
1 GbE	~1998
10 GbE	~2002
25 GbE	~2016
40 GbE	~2010
50 GbE	~2018
100 GbE	~2011
200 GbE	~2017
400GbE	~2017

option to implement software-defined infrastructure (SDI), a common platform to deliver the computing, network, and storage tasks. DPDK and Intel® server platform are great recipes to build out the software-defined network or security infrastructure and service. The early DPDK started from a performance test case of Intel® NIC. Today, DPDK software is widely recognized and deployed in the cloud and telecom network infrastructure. DPDK is an important software library to realize the NFV; it helps the concept; it helps in the production deployment.

NETWORK PROCESSOR

In the context of the hardware platform for the networking workload, it is necessary to discuss the network processor first. The telecom vendors have used network processors or similar chip technology as the primary silicon choice for data plane processing. Intel® was a market leader in this domain. Intel® had a product line known as Intel® Internet Exchange Architecture (IXA); the silicon product is known as IXP4xx, IXP12xx, IXP24xx/23xx, and IXP28xx. The technology was successful, and the product was a market leader. Within IXP silicon, a large number of microengines are inside of the processors; they are programmable for data plane processing. IXP has the XScale processor for the control plane, and it is a StrongARM-based silicon component.

In 2006, AMD has gained the processor leadership in x86 processor domain, and Intel® has to optimize the research and development organization, and investment portfolio. The network processor business was evaluated, and the internal finding indicated that IXP's overall business potential was not large enough for the long-term investment. Intel® gained the No.1 market share in 2006, but the market size was not big enough to support the long-term growth. Without the specialized network processor silicon, it requires other solution to address high-performance packet processing workloads. Intel®'s architects predicted that multicore x86 processor will be developed in a faster pace, so it makes sense to replace IXP roadmap with CPU-based technology. As a result, Intel® stopped the IXP product line development and gradually shifted towards CPU-based solution, which requires a software approach for networking workloads. This is an important strategic shift. The business plan is to converge all the networking-related solutions into x86-based multicore processor. The data plane workload requirement is different than the general computing characteristic; it needs to be fulfilled with the dedicated software solution; DPDK is a solution to respond to this strategic shift. The IXP is still existing at Intel®, which contributes to the accelerator technology, known as Intel® QuickAssist Technology (QAT), which is commonly available in QAT PCIe card, server chipset, or SoC (system on chip).

In 2006, the networking system needs to support the 10Gbps I/O. At that time, Linux system and kernel drivers were not able to achieve this. DPDK comes up as the initial software-based solution, as it meets up a new silicon trend at the dawn of multiple-core processor architecture. Since its birth, DPDK is very busy keeping up to the growing high-speed I/O such as 25Gbe, 40Gbe, and 100Gbe Ethernet.

In a short summary, DPDK was born in a disruptive time. The software design focuses on performance and scalability, and it is achieved by using multicores. Together with Intel's tick-tock silicon model, it sets the rapid product cadence compared to most of the network silicon technologies. Under the tick-tock model, Intel® released the new processor in such a cadence.

- CPU architecture needs to be refreshed every 2 years.
- CPU manufacturing process needs to be refreshed every 2 years.

At the early time, this is a very aggressive product beat rate in the silicon technology sector. Later, the smartphone silicon has the more aggressive schedule.

DPDK HISTORY

A network processor supports the packet movement in and out of the system at the line rate. For example, the system receives the 64-byte packet at a line rate of 10 Gbps, which is about 14.88 Mpps (million packets per second). This cannot be achieved in the early Linux kernel using x86 server platform. Intel's team started with a NIC performance test code; a breakthrough is made with NIC poll mode driver in Linux user mode. Traditionally, the NIC driver runs in Linux kernel mode and wakes up the system for the packet processing via interrupts, for every incoming packet.

In early days, CPU is faster than the I/O processing unit; the interrupt-based processing is very effective; CPU resource is expensive, so it is shared by different I/O and computing tasks. However, the processor and high-speed network interface speed mandates the packet processing to support 10 Gbps and above, which exceeds what the traditional Linux networking software stack can deliver. The CPU frequency still remains at 3 GHz or lower; only the game computer platform can do the frequency overclock up to 5 GHz. The networking and communication systems need to consider the energy efficiency as it is always on; it runs 24 × 7 hours. Therefore, the network infrastructure need to take the power consumption into the total cost of ownership (TCO) analysis. Today, most network systems run below 2.5 GHz for energy saving. From the silicon perspective, I/O speed is 10×~100× faster than before; CPU core count on a given platform is also up to 100×. One obvious way is to assign the dedicated cores to poll on the high-speed network ports/queues, so that the software can take advantage of the multicore architecture; this concept is a design foundation on DPDK.

At the very beginning, Intel® only shared the early prototype source code with the limited customers. Intel® shipped the early software package with FreeBSD license. 6wind played an important role to help in the software development and enhancement; there is a business contract. From 2011, 6wind, Wind River, Tieto, and Radisys announced business services support for Intel® DPDK. Later, Intel® shared DPDK code package on its website; it's free for more developers to download and use. 6wind set up the open-source website www.dpdk.org in April 2013; this became a host website; and eventually, DPDK becomes one of Linux Foundation Projects.

OPEN SOURCE

Today, any developer can submit source code patches via www.dpdk.org. At the beginning, DPDK focused on Intel® server platform enabling such as the optimal use of Intel® processor, chipset, accelerator, and NIC. This project grew significantly with the broad participation with many other silicon companies; this transformed DPDK to be ready with multiple architecture and multiple I/O (such as NIC, FPGA (field-programmable gate array)) support. Intel® is a foundational member to invest and grow the software community, together with other member companies and individual developers such as 6Wind, Redhat, Mellanox, ARM, Microsoft, Cisco, VMware, and Marvell (Cavium).

DPDK version 2.1 was released in August 2015. All major NIC manufacturers joined the DPDK community to release the NIC PMD support, including Broadcom NIC (acquired by Emulex), Mellanox, Chelsio, and Cisco. Beyond the NIC driver, DPDK is expanded for the packet-related acceleration technology; Intel® submitted the software modules to enable Intel® QAT for crypto acceleration, which is used for packet encryption/decryption and data compression.

The DPDK community has made great progress on the multiple architecture support. Dr. Zhu Chao at IBM Research China started the migrating DPDK to support Power Architecture. Freescale China's developers joined the code contribution. Engineers from Tilera and EZchip have spent efforts to make DPDK running on the tile architecture. DPDK also supported the ARM architecture later.

DPDK became a Linux Foundation Project in 2017: "The first release of DPDK open-source code came out 8 years ago; since that time, we've built a vibrant community around the DPDK project". Jim St. Leger, DPDK board chair, Intel®, wrote this statement "We've created a series of global DPDK Summit events where the community developers and code consumers gather. The growth in the number of code contributions, participating companies, and developers working on the project continues to reflect the robust, healthy community that the DPDK project is today". RedHat integrated DPDK into Fedora Linux first, then added to RedHat Enterprise Linux; many other Linux OS distribution packages followed this. VMware engineers joined the DPDK community and took charge of the maintainer role of VMXNET3-PMD, i.e., the de-facto high-performance software virtual interface to the guest software on VMware NSX. Canonical added DPDK support since Ubuntu 15. For public cloud computing, netvsc PMD can be used for Microsoft Hyper-V, and ENA PMD is available to support AWS Elastic Network Adapter.

EXPANSION

DPDK is designed to run in Linux user space; it is intended to stay closer to the application, which usually runs in Linux user space. The DPDK test case indicated a single Intel® core can forward packets with an approximate speed of 57 Mpps; this is achieved in an extremely simplified test case. Open vSwitch is a classical open-source component which is used by cloud infrastructure server. OVS integrated DPDK to accelerate virtual switching performance; this is widely adopted by the large-scale cloud computing and network virtualization systems, DPDK added the virtio–user interface to connect container with OVS-DPDK, and this is an overlay networking acceleration in a vendor neutral way.

For telecom equipment manufacturers, the server platform and open-source software, such as Linux, DPDK/VPP/Hyperscan, are important and new recipes to design and deliver the networking and security systems; they are also important recipes for cloud service model. Furthermore, Linux networking stack is also innovating fast with XDP and AF_XDP. This adds the interesting dynamics now as it offers the bypass path for Linux kernel stack, and it returns NIC management to the existing Linux utilities. It provides the new way to use Linux and DPDK together.

As one of the best open-source networking projects in the past decade, DPDK became a widely adopted software library for accelerating packet processing performance (10x more than Linux kernel networking) on the general-purpose server platforms. It is heavily used in many use cases, which are as follows:

- Build the network and security appliances and systems.
- Optimize the virtual switching for cloud infrastructure.
- Optimize storage systems with high I/O demand, like NVM device.
- NFV, build the software-centric networking infrastructure with servers.
- Cloud networking and security function as a service.

BOOK CONTRIBUTION

This book is a joint contribution from many individuals who worked and are working for Intel®. The early Chinese book editions are mainly contributed by Cunming Liang, Xuekun Hu, Waterman Cao (Huawei), and Heqing Zhu as the main editors. Each chapter in this book has the following list of contributors.

Section 1: DPDK foundation
Chapter 1: Heqing Zhu, Cunming Liang
Chapter 2: Chen Jing (Alibaba), Heqing Zhu
Chapter 3: Qun Wan, Heqing Zhu, Zhihong Wang

Chapter 4: Frank Liu (Netease), Heqing Zhu
Chapter 5: Yipeng Wang, Zhe Tao (Huawei), Liang Ma, Heqing Zhu
Chapter 6: Cunming Liang, Hunt David, Heqing Zhu
Chapter 7: Helin Zhang, Heqing Zhu
Chapter 8: Jingjing Wu, Xiaolong Ye, Heqing Zhu
Chapter 9: Wenzhuo Lu, Heqing Zhu
Chapter 10: Fan Zhang (Roy), Heqing Zhu
Section 2: I/O Virtualization
Chapter 11: Qian Xu, Rashmin Patel
Chapter 12: Tiwei Bie, Changchun Ouyang (Huawei), Heqing Zhu
Chapter 13: Tiwei Bie, Heqing Zhu
Section 3: DPDK Application
Chapter 14: Xuekun Hu, Waternan Cao (Huawei), Heqing Zhu
Chapter 15: Loftus Ciara, Xuekun Hu, Heqing Zhu
Chapter 16: Ziye Yang, Heqing Zhu

For the DPDK Chinese edition, the draft received the review and feedback from the below volunteers:

- Professor Bei Hua (USTC) and Kai Zhang (USTC, in Fudan University now);
- Professor Yu Chen, Dan Li (Tsinghua University);
- Dr. Liang Ou (China Telecom).

Lots of folks's work lead to this book content:

Intel®: Yong Liu, Tao Yang, De Yu, Qihua Dai, Cunyin Chang and Changpeng Liu, St Leger Jim, Yigit, Ferruh, Cristian Dumitrescu, Gilmore Walter, O'Driscoll Tim, Kinsella Ray, Konstantin Ananyev, Doherty Declan; Bruce Richardson, Keith Wiles, DiGiglio, John; Liang-min Wang, Jayakumar, Muthurajan Alibaba: Xun Li, Huawei Xie. This book leveraged the content from Redhat and VMware open-source developers and product managers. VMware: William Tu, Justin Petit; Redhat: Kim Buck, Anita Tragler, and Franck Baudin.

Special thanks to John McManara, Lin Zhou, Jokul Li, Ahern, Brian, Michael Hennessy Xiaomei Zhou, and Labatte Timmy for providing leadership support. Dan Luo/Lin Li helped in the translation and review from Chinese edition to the English version. Roy Zhang is very instrumental to guide the technology review of English edition.

DPDK is contributed by worldwide talents, who created the technology and made it thrive. At Intel®, it was mainly led by Intel® fellow: Venky (passed away at 2018) in Oregon and St leger Jim in Arizona.

Heqing Zhu

Editor

Heqing Zhu was born in China. He has worked with Intel® for 15 years. His roles include software developer, engineering leadership, product management, solution architect in telecom and cloud networking, and open-source software development. Prior to Intel®, he worked for Alcatel Shanghai Bell and Huawei. Mr. Zhu currently lives in Chandler, Arizona, in the United States. He graduated from the University of Electronic Science and Technology of China (UESTC) with a master's degree in Information and Communication System.

Contributors

Tiwei Bie
Ant Financial
Shanghai
China

Waterman Cao
Huawei
Shanghai
China

Jing Chen
Alibaba
Shanghai
China

Xuekun Hu
Intel®
Shanghai
China

David Hunt
Intel®
Shannon
Ireland

Cunming Liang
Intel®
Shanghai
China

Frank Liu
NetEase
Hangzhou
China

Jijiang(Frank) Liu
NetEase
Hangzhou
China

Ciara Loftus
Intel®
Shannon
Ireland

Wenzhuo Lu
Intel®
Shanghai
China

Liang Ma
Intel®
Shannon
Ireland

Changchun Ouyang
Huawei
Shanghai
China

Rashmin Patel
Intel®
Arizona
USA

Jasvinder Singh
Intel®
Shannon
Ireland

Qun Wan
Intel®
Shanghai
China

Yipeng Wang
Intel®
Oregon
USA

Zhihong Wang
Intel®
Shanghai
China

Jingjing Wu
Intel®
Shanghai
China

Qian Xu
Intel®
Shanghai
China

Ziye Yang
Intel®
Shanghai
China

Xiaolong Ye
Intel®
Shanghai
China

Fan(Roy) Zhang
Intel®
Shannon
Ireland

Helin Zhang
Intel®
Shanghai
China

Tao Zhe
Huawei
Shanghai
China

Section 1

DPDK Foundation

There are ten chapters in this section, focusing on the basic concepts and software libraries, including CPU scheduler, multicore usage, cache/memory management, data synchronization, PMD (poll mode driver) and NIC-related features, and software API (application programming interface). PMD is very important concept and the new user space software driver for NIC. Understanding the DPDK (Data Plane Development Kit) basics will build a solid foundation before starting on the actual projects.

In the first five chapters, we will introduce the server platform basics such as cache use, parallel computing, data synchronization, data movement, and packet forwarding models and algorithms. Chapter 1 will introduce the networking technology evolution, and the network function appliance trend, going from hardware purpose-built box to software-defined infrastructure, which is present in the cloud. Essentially, it is the silicon advancement that drives the birth of DPDK, and a few basic examples are given here. Chapter 2 will introduce memory and cache in the performance optimization context, how to use cache and memory wisely, the concept of the HugePage and NUMA (non-uniform memory access), and the cache alignment data structure. Chapters 3 and 4 will focus on multicore and multi-thread, the effective model for the data sharing for high parallelism, and the lock-free mechanism. Chapter 5 will move to the packet forwarding models and algorithms, and the decision is required to choose the run-to-completion and/or pipeline model, or both.

The next five chapters will focus on I/O optimization, and we will talk about PCIe, NIC, and PMD design and optimization details, which enables DPDK PMD to deliver the high-speed forwarding rate to meet demands at 10 Gbe, 25 Gbe, and 40 Gbe up to 100 Gbe in a server platform. Chapter 6 will have a deep dive on PCIe transaction details for the packet movement. Chapter 7 will focus on NIC performance tuning, and the platform and NIC configuration. Chapters 8 and 9 will go further into NIC common features and software usage, the multi-queue, flow classification, core assignment, and load balancing methods to enable a highly scalable I/O throughput, and will introduce the NIC offload feature on L2/L3/L4 packet processing. Chapter 10 is about packet security and crypto processing; securing the data in transit is an essential part of Internet security. How can DPDK add value to "encrypt everywhere"?

1 Introduction

Heqing Zhu and Cunming Liang
Intel®

CONTENTS

1.1   WHAT'S PACKET PROCESSING?

Depending on whether the system is a network endpoint or middlebox, packet processing (networking) may have different scope. In general, it consists of packet reception and transmission, packet header parsing, packet modification, and forwarding. It occurs at multiple protocol layers.

- In the endpoint system, the packet is sent to the local application for further processing. Packet encryption and decryption, or tunnel overlay, may be part of the packet processing, session establishment, and termination.
- In the middlebox system, the packet is forwarded to the next hop in the network. Usually, this system handles a large number of packets in/out of the system, packet lookup, access control, and quality of service (QoS).

The packet may go through the hardware components such as I/O (NIC) interface, bus interconnect (PCIe), memory, and processor; sometimes, it may go through the hardware accelerator in the system. Most of the packet movements and modifications can be categorized as follows:

- Data movement is, like packet I/O, from PCIe-based NIC device to cache/memory, so that CPU can process further.
- Table lookup/update, this is memory access (read/write), this is used for packet-based access control, or routing decision (which interface to be sent out).
- Packet modification involves network protocols that are defined at many different layers, just like peeling the onion layers; each protocol layer has its data format, and usually, it is defined by the International Standard like Internet Engineering Task Force/ Request for Comments (IETF/RFC). Packet processing often involves the packet change, header removal, or addition.

1.2   THE HARDWARE LANDSCAPE

Traditionally, the network system is highly complicated and consists of the control, data, signal, and application plane; each plane can be realized with the different subsystems; and these systems are known as the embedded systems with low power consumption, low memory footprint, but real-time characteristics. Such systems require the hardware and software talents to work together.

In early 2000, CPU only had a single core with high frequency; the first dual-core processor for general computing emerged in 2004. Prior to that, the multicore, multithread architecture is available in the networking silicon, but not in the general-purpose processors. In the early years, x86 was not the preferred choice for packet processing. As of today, the below silicon can be used for packet processing system. From the programmer skills, they can be split into the different category.

- Hardware accelerator (FPGA (field-programmable gate array), ASIC (application-specific integrated circuit));
- Network processor unit (NPU);
- Multicore general-purpose processor (x86).

These systems are used for different scenarios; each hardware has certain advantages and disadvantages. For large-scale and fixed function systems, the hardware accelerator is preferred due to its high performance and low cost. The network processor provides the programmable packet processing, thereby striking a balance between flexibility and high performance, but the programming language is vendor specific. In the recent years, P4 has emerged as a new programming language for packet processing, and it gained the support from Barefoot Switch and/or FPGA silicon, but not common for NPU.

The multicore general-purpose processor has the traditional advantages such as supporting all generic workloads and the server platform that is commonly equipped with high-speed Ethernet adapters. The server has quickly evolved as the preferred platform for packet processing. It can support the complex packet processing together with the application and service; the application and service software can be written with many different programming languages (C, Java, Go, Python). Over the years, there are lots of high-quality open-source projects that emerged for packet processing, such as DPDK, FD.io, OPNFV, and Tungsten.io. The cloud infrastructure has gone down a path known as NetDevOps approach, taking the open source to deliver software-defined networking and security infrastructure and service.

From the perspective of the silicon advancement, new accelerator and high-speed I/O units have been integrated with multicore processors. This leads to the generation of system on chip (SoC). SoC is cost-effective. Silicon design has longer life cycles.

1.2.1 HARDWARE ACCELERATOR

ASIC and FPGA have been widely used in packet processing. Hardware developers are required to implement the chip and use the chip. An ASIC is an integrated circuit designed for special purpose. This integrated circuit is designed and manufactured based on the specific requirements of target systems. ASIC is designed for specific users' needs; it needs the large-volume production to afford the high R&D cost; it is smaller in size; and it has lower power consumption, high reliability and performance, and reduced cost, in comparison with the general-purpose processor. ASIC's shortcomings are also obvious: not flexible, not scalable, high development costs, and long development cycles. ASIC leads to the development of the popular accelerators such as crypto and signal processing. Combining ASIC with the general-purpose processors will lead into SoC that provides heterogeneous processing capability. In general, the dedicated board design is needed to use ASIC.

FPGA is a semi-custom circuit in the ASIC domain. Unlike ASIC, FPGA is programmable and flexible to use. FPGA is inherently parallelized. Its development method greatly differs from the software. FPGA developers require an in-depth understanding of hardware description language (Verilog or VHDL). The general software executes in the sequential order on the general-purpose processor; the software parallelism is up to the software design. FPGA silicon can include some fixed functions and some programmable functions. FPGA has made great progress in the data center in the recent years, and FPGA can be used as a cloud service. FPGA can be used for smart NIC. FPGA is often selected to build the super-high-speed I/O interface, advanced packet parsing and flow filtering, and QoS acceleration. FPGA can be offered as add-in card; through PCIe interface, it is easy to be plugged into the server system; it is popular for cloud data center scenario. FPGA can also be used for a specific purpose, like signal processing. FPGA is also often used in a special board design.

Take the 5G wireless base station as an example; the telecom vendor develops the system in stages, and it may use FPGA to build the early-stage product. Once the product quality is in good shape, the high-volume needs will drive the new stage, which focuses on using ASIC (SoC) to replace FPGA, which will drive the cost down for the large-scale use.

1.2.2 Network Processor Unit

NPU is a programmable chip specifically designed for packet processing. It is usually designed with a multicore-based parallel execution logic, dedicated modules for packet I/O, protocol analysis, routing table lookup, voice/data encoding/decoding, access control, QoS, etc. NPU is programmable, but not easy; the developer needs to take a deep dive into the chip's datasheet and learn the vendor-specific instruction sets, known as microcode (firmware); the user needs to develop the hardware-based processing pipeline for the target network application. The network applications are realized with the loadable microcode (firmware) running on NPU. NPU generally has the built-in high-speed bus and I/O interface technology. In general, NPU has the built-in low latency memory modules; it allows the forwarding table using the on-chip memory, which makes it faster than the external DRAM access. NPU can be integrated as part of SoC; in recent years, NPU-based silicon vendors have been consolidated by CPU or NIC vendors

The below diagram (Figure 1.1) is a conceptual diagram. The "packet processing engines" is programmable hardware logic, which allows the rapid implementation of workload-specific packet processing; the written microcode can run on many parallel engines. "Physical I/O" interface is a fixed function to comply with the standardized interface specification. "Traffic manager" and "classification and queueing" are relatively fixed functions, which are common features in most of the network systems. They are built in as the specialized hardware units for QoS and packet ordering. "Internal memory" and "TCAM (ternary content-addressable memory)" provide the low latency access memory for the packet header parsing and forwarding decision. The memory controller connects the external memory chips for larger memory capacity.

NPU has many advantages, such as high performance and programmability, but its cost and workload-specific characteristics imply the market limit. It is often used in communication systems. Different NPU products have been available from various silicon vendors. As said before, NPU microcode is often vendor-specific; thus, it is not easy to use, and it is not easy to hire the talented developers who understand NPU well. Due to the steep learning curve, the time to market will be affected by the availability of talents. Because NPU is used to its limited market, it does not create enough job opportunities. The experienced talents may shift focus to leave the network processor domain to seek the career growth opportunities elsewhere.

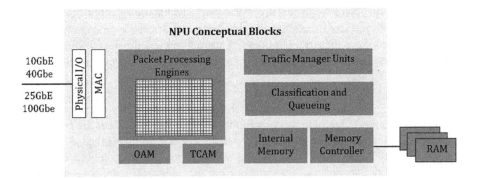

FIGURE 1.1 NPU conceptual block.

There are many attempts of using NPU with the common programming languages (like C); technically speaking, it is possible and it is available, but it is not the best way of using NPU. But the other reality is the performance gap between using C and using microcode language. Translation from C programming language to microcode is doable, but it is not the optimal way of getting the most performance out of NPU. So this path is feasible, but does not have the real value. If all NPUs from different vendors can support P4 language, it will be much easier to use. Though, the performance dilemma may remain as a similar concern. P4 ecosystem is still in its development phase, and the broad NPU/P4 support is yet to be seen.

1.2.3 MULTICORE PROCESSOR

In the past 15 years, CPU delivered huge boost on the processing cycles, thanks to the new era of the multicore architecture. With more cores available in the general-purpose processor, it is natural to assign some cores for the packet processing purpose; CPU finds its way to converge more workload on its own. From the historic networking system perspective, (low-power) CPU has always been preferred for control plane (protocol handling), and some network services (like wireless data service) are computer intensive. A few cores can be assigned to handle the data plane processing, whereas the remaining cores can be used for control and service plane processing, which can make a more efficient and cost-effective system.

Looking back, take a telecom appliance in 2005, which is a complicated system (chassis) consisting of ten separate boards; each board is a subsystem having a specific function, and each board is built with a dedicated silicon and software. So different boards within a single system have different platform designs and silicon components. Later, the idea is to converge all the subsystems into a powerful server system, and the subsystems can run as software threads (within virtual machines or containers). This new approach is expected to transform the network industry with less silicon components, less heterogeneous system and platform architecture, and hence less cost. This is the initial motivation for implementing network functions with software.

On the other hand, CPU release cycle is short, and new processors come to the market on a yearly basis. But the network processors and accelerators are difficult to catch up with this fast release cadence, which often happens every 3 or 4 years. The market requires the large-scale shipping of the general-purpose processor, and the market does not have the strong business demand to refresh the networking silicon. Over the years, the business drives the significant progress on the general-purpose processor with a very competitive cost.

Figure 1.2 describes a dual-socket server platform. Two Xeon processors are interconnected with UPI (Ultra Path Interconnect) system bus; memory channels are directly connected to both processor units; all the external devices are connected via the PCIe interface; each socket is connected with 2×25 Gbe Ethernet adapters (NIC) with 100 Gbps I/O for data in and out. Lewisburg PCH is a chipset, which is served as the platform controller, supporting the additional management engine, such as high-speed I/O (USB, SATA, PCIe), 4×10 Gbe Ethernet interface, and Intel® QAT (built-in security accelerator for crypto and compression functions). The processor, memory, and Ethernet devices are the main hardware modules to handle the packet processing tasks. PCI-e interface provides the I/O extensibility and supports to add the differentiated accelerators (or more flexible I/O) into server platforms, e.g., FPGA cards

The general-purpose processor can be integrated with additional silicon IP; then, it will evolve into SoC. The SoC system often consists of a processor, the integrated memory controller, the network I/O modules, and even hardware accelerators such as security engine and FPGA. Here are a few known SoC examples:

- *Intel®*: Xeon-D SoC, Atom SoC;
- *Tilera*: TILE-Gx;
- *Cavium network*: OCTEON & OCTEON II;
- *Freescale*: QorIQ;
- *NetLogic*: XLP.

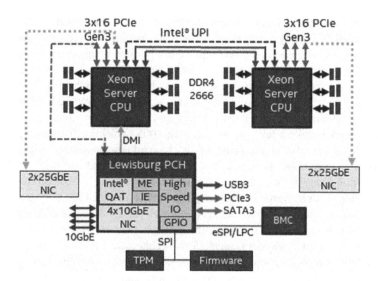

FIGURE 1.2   A dual-socket server platform.

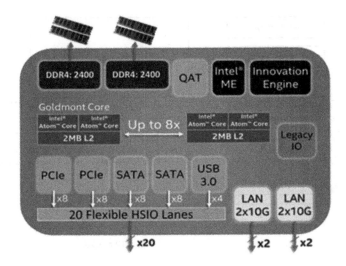

FIGURE 1.3   Intel® Atom SoC diagram.

The block diagram as shown in Figure 1.3 is one example of Intel® Atom SoC, which is a power-efficient silicon; it's the tiny chip consists of 2x atom core, internal chipset, Intel® QAT, and 4 × 10 Gbe Ethernet interface. Given that all the hardware units are integrated into a single chip, SoC is viewed as power efficient and cost-effective. The power consumption of this chip is less than 10 W. DPDK can run on this chip to move the packet data from Ethernet to CPU with the line rate of 10 Gbps. SoC is also popular in the ARM-based processor design. For example, AWS released the Graviton SoC for cloud computing use case. It is important to point out that CPU has made huge progress with cloud virtualization and container technologies, which provides more granularity and flexibility to place software workload with computation resources. CPU-based packet processing reduces the needs for hardware talents; it deals mostly with the software.

1.3   THE SOFTWARE LANDSCAPE

DPDK is created for high-speed networking (packet processing). Before DPDK was born, most of the personal computers and server systems were installed with Windows or Linux OS. All these systems had the networking capabilities, and support the network protocol stack and talk to each other via Ethernet and socket interface; the low-speed network processing capability is good enough for a computation-centric system. There is a big difference in supporting low-speed network processing (10/100 Mbps) and supporting high-speed network processing (1/10/25/40/50/100 Gbps). Before the multicore architecture was common in a CPU, it was not a popular idea to use an Intel® processor for network processing system.

Let's take a look at the popular wireline and wireless network protocol stacks, and a little more specific on what're the network protocol layers. Figure 1.4 shows an example of classic wireline network protocol layers (including both OSI model and TCP/IP model). The left side shows the OSI 7-layer model, whereas the right

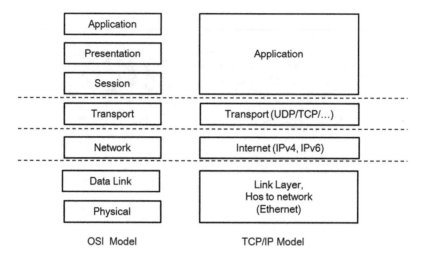

FIGURE 1.4   The OSI model and TCP/IP model (wireline).

side shows the TCP/IP model. The TCP/IP model is often implemented by the Linux kernel systems. By default, the incoming packet goes to the Linux kernel at the "link layer", and this is often done by NIC and its driver, which is running in the kernel space. The whole network stack can be handled by the Linux kernel. In the network system, the packet has to be copied from the kernel space to the user space because the application usually resides in the user space, and this application will eventually consume the arrived packet. Before zero copy is introduced to Linux stack, packet copy is an expensive processing, but it is essential, as packet is received in Linux kernel but consumed by the user space application. For the middlebox networking systems (like routers), routing functions are largely implemented in the user space as well, not using the Linux kernel space stack. Generally, software development in the kernel space is much harder in terms of debugging and testing. User space software is easy to develop/debug. There are certain systems to handle everything in the kernel, but they are rare.

Figure 1.5 describes wireless 4G (LTE) user plane network protocol layers. eNo-deB is a wireless base station with air interface. Serving gateway (GW) and PDN GW are the wireless core networking systems. For the 5G system, the user plane stack is very similar to 4G, eNodeB becomes gNodeB, and serving GW/PDN GW becomes UPF. As we can see, the protocol layers are different between the base station and the core system. In the wireless base station (eNodeB), L1/L2 protocols are different because the wireless interface is essentially air interface and uses wireless signal processing and codec technology. We will not cover any details here. It is a completely different technology domain. From eNodeB to serving GW, the packet processing system is similar to a typical wireline network system, where L1/L2 is based on Ethernet interface and GTP/UDP/IP protocols are running on the top of Ethernet. 5G network stack is also similar. 5G prefers to use cloud architecture in order to implement the service-oriented elastic model, so that the edge computing

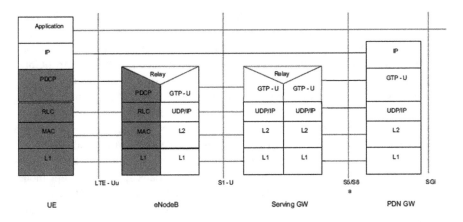

FIGURE 1.5   Wireless user plane network protocol (4G/LTE).

and network slicing are part of the new service. Packet processing and computing service are converged at the 5G network infrastructure node. This is largely due to the multicore CPU that has the huge computation power. In the early years, not a long time ago, the computing system focused on singular workload, or singular service provision, where the packet processing requirement is low and the Linux kernel stack is sufficient to handle less than 1 Gbps Ethernet interface. Later, the networking systems need support multiple high-speed networking interfaces (far more than 1 Gbps), and they require different software options. Linux kernel networking cannot meet the higher networking demand.

1.3.1   Before DPDK

In the early 2000s, Intel® processor was not widely used for high-speed network processing. NPU was a silicon choice at Intel®; now the change must happen; a pathfinding effort at Intel® is kicked off.

How does a traditional NIC device process the packet in a server system using Linux? The steps are summarized below:

- A packet arrives at a NIC (PCIe device).
- The NIC completes DMA (direct memory access) and copy packet into host memory region known as a packet buffer.
- The NIC sends an interrupt to wake up the processor.
- The processor reads and writes the packet descriptor and packet buffer.
- The packet is sent to the Linux kernel protocol stack for more protocol processing like IP-related access control decision.
- If the application resides in the user space, the packet data (also known as payload) will be copied from the kernel space to the user space.
- If the application resides in the kernel space, the data will be processed in the kernel mode (less percentage).

In the early system, each incoming packet may trigger an interrupt. The interrupt overhead includes the context switching, and it is affordable if there are not many packets coming to the system in a short period. In the past decade, CPU frequency remained almost the same, but Ethernet speed jumped from 10/100 Mbps to 1 Gbps, 10 Gbps to 25 Gbps, and 40 Gbps to 100 Gbps. Software is facing the challenge to handle the large packet burst scenario on a growing network interface, huge number of packets will arrive, and the system cannot afford high amount of interrupt processing. Simply speaking, the overhead is too high. NAPI (new API) mechanism was introduced into the Linux kernel. It allowed the system, after wakeup by the interrupt, to initiate the software routine that processes multiple packets in a polling manner, until all packets are handled, then goes back to the interrupt mode. The NAPI method can significantly improve the packet processing efficiency in a high burst scenario. Later, Netmap (2011), a well-known high-performance network I/O framework, uses a shared pool of packets to reduce the packet replication from the kernel space to the user space. This solves the other problem—the high cost of packet copy [1,2].

Netmap and NAPI have significantly improved the packet processing capability on legacy Linux system. Is there any further improvement room? As a time-sharing operating system, Linux will schedule many tasks with time-slicing mechanism. Compared with equal time assigned to all tasks, Linux scheduler has the job to assign the different time slices for different tasks. The number of CPU cores was relatively small in earlier years, and in order to allow every task to be processed in a timely fashion, time sharing was a good strategy to support multiple tasks to sharing the expensive processor cycles, although this method was done at the cost of efficiency. Later, CPU has more cores available, so it is time to look at the new ways to optimize the system performance. If the goal is to pursue high performance, time sharing is not the best option. One new idea is to assign the dedicated cores to the dedicated tasks. Netmap reduces the memory copy from the kernel space to the user space, but there are still Linux schedulers, which do not eliminate the overhead of the task switch. The additional overhead from the task switch and the subsequent cache replacement caused by the task switch (each task has its own data occupied in cache) will also have an adverse impact on the system performance.

By nature, network workload is latency sensitive, and packet may traverse many hops through Internet; as a result, real time is a critical system requirement for the networking system. It is a long processing path from the NIC interrupt to the software interrupt routine (served by CPU), and then the packet payload is handled by the final application; this path takes lots of cycles.

Prior to 2010, x86 CPU was not a popular silicon choice to design the high-speed packet processing system. In order to complete the solution transition from NPU-based silicon to x86/software, there are a few fundamental challenges. Intel® engineers need answer for the following:

- A software path to enable the packet processing on x86 CPU;
- Find a better software method to do things differently;
- A performance scale way using multicore architecture;
- How to tune "Linux system" as packet processing environment.

1.3.2   DPDK Way

DPDK is the answer to the above challenges; particularly, PMD (polling mode driver) has been proved as a high-speed packet processing software library on Linux. DPDK, essentially, is based on a set of software optimization principles, a set of software libraries to implement the high-performance packet movement on the multicore processor. The initial goal is to focus on the high-speed packet I/O on the server platform. This software demonstrated that the server is good for the networking system, and it can handle the high-speed data plane. The journey was not easy, and it was achieved through heavy engineering investment. It is built on the many software optimization practices. Let's navigate a few ideas and techniques quickly.

Polling mode: Assign the dedicated core for NIC packet reception and transmission. This approach does not share core for other software tasks, and the core can run in the endless loop to check if any packet just arrives or needs to be sent out, thus reducing the need for interrupt service and its overhead. We will discuss the trade-offs between polling and interrupt mechanisms later. In fact, DPDK supports both mechanisms and even hybrid-use model.

User space driver: In fact, in most scenarios, the packet needs to be sent to the user space eventually. Linux NIC driver is mostly kernel based. The user space driver can avoid unnecessary packet memory copy from the kernel space to the user space, and it also saves the cost of system calls. An indirect benefit is that the user space driver is not limited to the packet buffer structure mandated in the Linux kernel mode. Linux kernel stack mandates the stable interface, and the DPDK-based mbuf (memory buffer) header format can be flexibly defined (because it is new) so that it can be designed in DMA-optimized way for NIC. This flexibility adds the performance benefit. The user space driver is flexible, it is easy to modify, and it meets the rapid development needed for different scenarios.

Core affinity: By setting a thread's affinity to a particular CPU core, specific tasks can be bound with cores (thread). Without the core affinity assignment, there might be the task switching among different cores, and the drawback to this assignment is that thread switching between cores can easily lead to performance losses due to cache misses and cache write-back. One further step is to ask a core to be excluded from Linux scheduling system, so that the core is only used for the specific task.

Optimized memory: Network processing is an I/O-bound workload scenario. Both CPU and NIC need access to the data in memory (actually cache and/or DRAM) frequently. The optimal memory access includes the use of HugePage and contiguous memory regions. For example, HugePage memory can reduce the TLB misses, multichannel-interleaved memory access can improve the total bandwidth efficiency, and the asymmetric memory access can reduce the access latency. The key idea is to get the data into cache as quickly as possible, so that CPU doesn't stall.

Software tuning: Tuning itself cannot be claimed as the best practice. In fact, it refers to a few known tuning practices, such as cache line alignment of data structure, avoiding false sharing between multiple cores, pre-fetching data in a timely manner, and bulk operations of multiple data (multi-buffer). These optimization methods are used in every corner of DPDK. The code example can be found in the "l3fwd" case study. It is important to know that these techniques are commonly applicable; beyond DPDK, any software can be optimized with the similar approach.

Using the latest instruction set and platform technologies: The latest instruction sets of Intel® processor and other new features has been one of the innovation sources of DPDK optimization. For example, Intel® DDIO (Direct Data I/O) technology is a hardware platform innovation in DMA and the cache subsystem. DDIO plays a significant role to boost I/O performance as the packet data can be directly placed into cache, thus reducing the CPU access latency on DRAM. Without DDIO, packet is always placed into memory first, and then CPU needs to fetch packet data from DRAM into cache, which means the extra cycles that CPU needs to wait. The other example is how to make the best use of SIMD (single-instruction multiple data) and multiple buffer (multi-buffer) programming techniques. Some instructions, like CMPXCHG, are the cornerstone for lockless data structure design. Crc32 instruction is also a good source for efficient hash computation. These contents will be covered in later chapters.

NIC driver tuning: When the packet enters the system memory through PCIe interface, I/O performance is affected by the transaction efficiency among the PCIe-based device, bus transaction, and the system memory. For example, the packet data coalescence can make a difference through transferring multiple packets together, thus allowing a more efficient use of PCIe bus transactions. Modern NICs also support load balancing mechanisms such as receive side scaling (RSS) and Flow Director (FDir) features, which enable NIC multiple queue to work with CPU multiple core model. New NIC offload can also perform the packet header checksum, TCP segmentation offload (TSO), and tunnel header processing. DPDK is designed to take full advantage of the NIC features for performance reasons. These contents will be described in Chapters 6–9.

Network virtualization and cloud-native acceleration: Initial DPDK optimization focuses on moving packets from I/O to CPU. Later, DPDK provides the optimal way to move packets from host to tenants (VM, container tenants). This is a crucial ingredient for cloud infrastructure and network function virtualization (NFV). DPDK supports both SR-IOV and vSwitch optimization with PMD concept.

Security acceleration: DPDK can run from the bare metal to the virtualized guest and container-based environment; the initial Application Programming Interface (API) abstraction is NIC centric; later, it is extended from Ethernet to crypto, compression, and storage I/O acceleration. Crypto and compression APIs are important software abstraction; they can hide the underlying silicon's implementation difference.

1.3.3 DPDK Scope

Here are the basic modules within DPDK. It mimics the most network functions in software and serves as a foundational layer to develop a packet processing system (Figure 1.6).

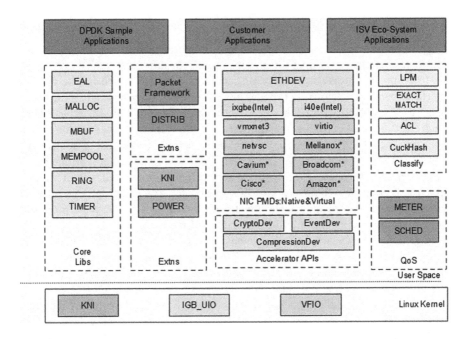

FIGURE 1.6   DPDK framework and modules.

Core libraries (core libs) provide the Linux system abstraction layer and provides software APIs to make use of the HugePage memory, cache pool, timers, lock-free rings, and other underlying components.

PMD libraries provide all user-space drivers in order to obtain a high network throughput by PMD. Basically, all industry-leading Ethernet companies are offering the PMD at DPDK. In addition, a variety of virtual NICs for Microsoft (netvsc) and VMware (vmxnet3), and KVM-based virtualized interfaces (virtio) are also implemented.

Classify libraries support exact match, longest prefix match (LPM), wildcard matching (ACL [access control list]), and cuckoo hash algorithm. They focus on flow lookup operations for common packet processing.

Accelerator APIs supported the packet security, data compression, and event modeler for core–core communications. The FPGA accelerated function or SoC units can be hidden under the abstract software layers here.

QoS libraries provide network QoS components such as Meter and Sched.

In addition to these components, DPDK also provides the platform features like POWER, which allows the CPU clock frequency to change at runtime for energy saving, and KNI (kernel network interface), which builds a fast channel to the Linux kernel stack. The Packet Framework and DISTRIB provide the underlying components for building a more complex multicore pipeline processing model. This is an incomplete picture as the DPDK project evolves quickly with new release in every quarter. In DPDK, most components are BSD (Berkeley Software Distribution)-based license, making it friendly for the further code modification and commercial adoption.

1.4 PERFORMANCE LIMIT

The performance limit can be estimated by the theoretical analysis. The theoretical limit can come from multiple dimensions. Take an example of packet processing; the first limit is the packet forwarding rate, which is determined by the physical interface speed, also known as the line speed on a given interface. When the packet enters the memory from NIC, it will go through the I/O bus (e.g., PCIe bus). There is a limit at the PCIe bus transaction level. Of course, there is a ceiling for CPU to load/store packet data to cache lines. For example, Intel® Haswell processor can only load 64 bytes and store 32 bytes in a cycle. The memory controller is limited by a memory read/write bandwidth. All these hardware platform boundaries in different dimensions contribute to the workload's performance limit. Through optimization, the software developer can look at these dimensions to write a high-performing software; the goal is to get closer to the performance limit.

In theory, I/O interface, PCIe bus, memory bandwidth, and cache utilization can set quantitative limits; it is easy to say but not easy to do, as it requires in-depth system-level understanding. Good developers know how to measure the performance tuning progress and find potential room for improvement. It takes a lot of effort to tune the workload and then get closer the theoretical limits. If the software is already extremely optimized, there will be no good return to continue pushing the boundary.

DPDK has a design goal to provide the software libraries so that we can push the performance limit of packet processing. Is it designed in a way that has already reached the system limit? As we try to gain a better understanding of DPDK, let's get back to the simple facts. What are the common performance metrics to measure the packet processing?

1.4.1 THE PERFORMANCE METRIC

The performance indicators for packet processing such as throughput, latency, packet loss, and jitter are the most common metrics. For packet forwarding, the throughput can be measured as packet rate—pps (packets per second); another way to measure is bit rate—bps (bits per second). Bit rate is often associated with the physical interface, like NIC port speed. Different packet size often indicates different requirements for packet storing and forwarding capabilities. Let's establish a basic concept of effective bandwidth and packet forwarding rates.

The line rate (transmitting in the wire speed) is the maximum packet (or frame) forward rate, which is limited by the speed of physical interface, theoretically. Take Ethernet as an example. The interface is defined as 1 Gbps, 10 Gbps, 25 Gbps, 40 Gbps, and 100 Gbps; each represents the maximum transmission rate, measured in bps. Indeed, not every bit is used to transmit the effective data. There is an inter-packet gap (IPG) between Ethernet frames. The default IPG is 12 bytes. Each frame also has a 7-byte preamble and a 1-byte Ethernet start frame delimiter (SFD). The Ethernet frame format is shown in Figure 1.7. The effective data in Ethernet frame mainly includes the destination MAC (media access control) address, source MAC address, Ethernet type, and payload. The packet tail is known as FCS (frame checksum) code, which is designed to validate the Ethernet frame integrity (Figure 1.7).

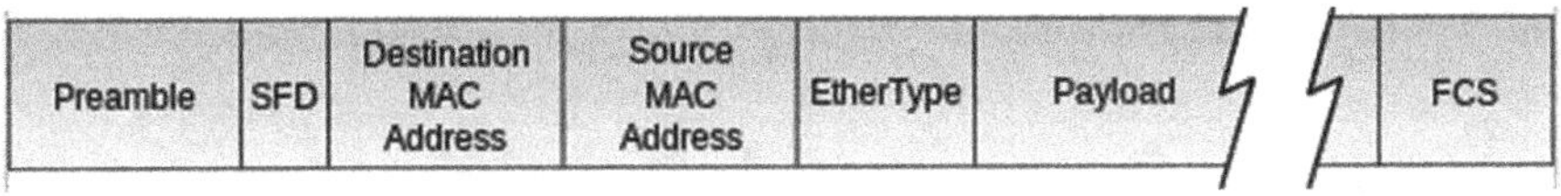

FIGURE 1.7   Ethernet frame format.

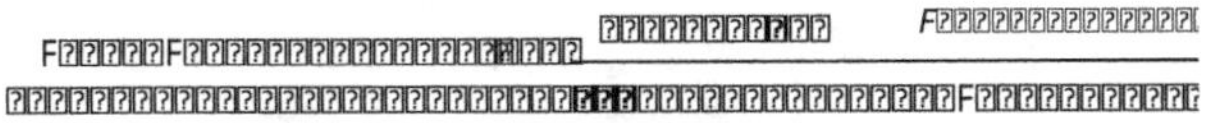

FIGURE 1.8   Packet forwarding rate: From bps to pps.

TABLE 1.1

Packet Forwarding Rate with Different Packet Sizes

Interface Speed	10 Gbps		25 Gbps		40 Gbps	
Packet size (Byte)	# of Mpps	arrival (ns)	# of Mpps	arrival (ns)	# of Mpps	arrival (ns)
64	14.88	67.20	37.20	26.88	59.52	16.80
128	8.45	118.40	21.11	47.36	33.78	29.60
256	4.53	220.80	11.32	88.32	18.12	55.20
512	2.35	425.60	5.87	170.24	9.40	106.40
1024	1.20	835.20	2.99	334.08	4.79	208.80

The Ethernet frame forwarding rate and the bit rate can be translated by the equation shown in Figure 1.8.

As shown in Table 1.1, if the data is transmitted in full interface speed, the small packet implies the high packet arrival rate. In a nutshell, small packets cause the larger processing burden as more packets will arrive in a given time, and it can be 10× more overhead for small packet (64 bytes) than for normal packet

1.4.2   THE PROCESSING BUDGET

What is the biggest challenge for processing packets on a general-purpose processor like CPU? Take 40 Gbps Ethernet as an example; the curve shown in Figure 1.9 indicates the maximum forwarding rate for the different packet sizes. Internet traffic is a mix of many packets of different sizes

For the packet size of 64 bytes and 1024 bytes, respectively, CPU instruction cycles (as the processing budget) are different to meet 40 Gbps forwarding requirement, i.e., 33 cycles vs 417 cycles. It is a very different system load when dealing with small or large size packets. If the packet size is smaller, the packet interval is shorter, i.e., 16.8 ns vs 208.8 ns. Assuming that the CPU frequency is running at 2 GHz, 64-byte and 1024-byte packets can, respectively, consume 33 and 417 clock cycles to reach the line rate. In the store-forward model, the packet receiving, transmitting, and forward table lookup need to access the memory. CPU can only wait if there is memory access; this access takes the cycles, known as memory latency. The actual

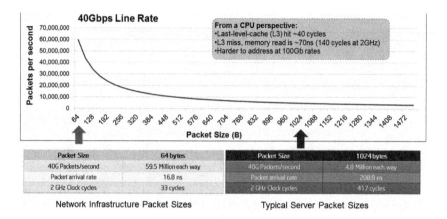

FIGURE 1.9   Cost of packet instructions at 40 Gbps line rate.

latency depends on where the data is located. If data resides in the last level cache (LLC), the access takes about 40 clock cycles. If there is an LLC miss, it will need an external memory read, with an estimated latency of 70 ns, which is translated to 140 cycles on a 2-GHz system. For a small packet size (64 bytes) to reach 40 Gbps, the total budget is just about 33 cycles. This poses a challenge to the general-purpose CPU system for high-performance network workload use. Does this rule out CPU for high-performance network workload? The answer is no.

DPDK has a solution to tackle the challenge; a set of optimization methods are implemented to make the best use of the hardware platform features (DDIO, HugePage memory, cache alignment, thread binding, NUMA (non-uniform memory access) awareness, memory interleave, lock-free data structure, data pre-fetching, use of SIMD instructions, etc.). By combining all these optimization methods, the memory latency challenge can be minimized. A single core can do L3 packet forwarding rate at about 40 Mpps; it is measured in a simplified test case, which just moves packets from Ethernet to CPU. Giving that there is an increasing number of CPU cores that are available, I/O performance can scale up with more Ethernet adapters that are inserted with more PCIe slots. The L3 forwarding throughput has been measured in 4-socket system, and it can reach about 1TPPS on Intel® Xeon Scalable Processor. It is a significant milestone announced by an open-source project—FD.io/VPP in 2017 [3].

DPDK is not yet able to handle the extreme high I/O use case. For example, the optical transport interface needs support from 400 Gbps interface. This is too much for a software-only approach on a server platform; the PCIe bus and NIC are not available yet. DPDK reduces the network system development barrier, and now software developers can implement the network functions.

1.5   DPDK USE CASE

DPDK is the new open-source technology, and it is rising together with the server platform, which is going through the rapid development, and has more CPU cores and higher-speed Ethernet interface. The server platform cost is very appealing when compared to the mainstream network/storage systems (which are expensive due to their purpose-built silicon-based system design). From business side, this open-source project reduces the software investment. Overall, the trend is to build the software-defined infrastructure on server platforms, and DPDK has been proved to deliver the accelerated network, and computing and storage functions.

1.5.1 ACCELERATED NETWORK

DPDK, as open source, can be of immediate use without license cost. What's more exciting, it delivers a performance boost with each new generation of server platform. Processor refresh means the improved IPC (instruction per cycle) at per core level, and it may be enhanced with a more effective system bus that is connected with more cores. DPDK on a single core can deliver 10/25 Gbps easily. Together with multi-queues in NIC, DPDK drives the 40/100 Gbps interface using two or more cores. Xeon processor has many cores, and the packet throughput bottleneck is often limited by PCI-e (I/O) interface and lanes. Indeed, I/O is the limit, not CPU.

Prior to DPDK (or similar technology), the network infrastructure system is often designed with a highly complicated system; hardware and software co-design is common; it requires the engineers to know both hardware and software skills. So the network system was mainly done by a large-sized engineering organization with high cost. It involves the system chassis; different but purpose-built boards are connected as the subsystems, which may meet the needs of signal processing, data plane, control, and application services. Each board is built with the specific embedded software and its function; hence, it requires the dedicated engineering investment.

Decoupling the hardware and software development, and building software on the top of common server system, is a big and disruptive change; the server platform and high-speed NIC are easy to get; it is flexible to add more NICs to meet more I/O demand; the cost is very transparent on the server platform. It is easy to find software talents who can write code, and load and debug the program on the standard server platform. It is difficult to find the designers who understand the NPU or Network ASIC/FPGA to build the highly complex system. From the cloud computing industry, the software developers adopt the open source on the server platform, so they built the load balancer and the anti-DDoS system using DPDK, and they run the software in a standard server for its in-house production deployment. This is the early example to replace the commercial load balancer system. Later, the network gateway can be moved to run within a virtual machine, and it can be offered as an elastic network service.

Similar idea known as network function virtualization (NFV) emerged in 2012. Server virtualization leads to the cloud computing; it allows multiple tenants to run workload on the logically isolated environment but consolidated on a physically shared server. By replicating this to the telecom network infrastructure, NFV is intended to run the network functions as a tenant workload and consolidate multiple workloads on a physically shared server. This will allow telecom service providers to get more flexibility in choosing the vendor and solution suppliers, e.g., decoupling the hardware and software suppliers. This idea is an important step to build out 5G infrastructure and network service. Software defined networking (SDN)/NFV is a big wave to transform network infrastructure; to build the new system, DPDK is valuable to be part of software architecture.

1.5.2 ACCELERATED COMPUTING

For network nodes, it is very easy to understand the value of DPDK. In fact, DPDK is very helpful for the cloud computing node, like Open vSwitch acceleration. Open vSwitch is responsible for creating the overlay network for multiple tenant's isolation needs; any traffic to tenant workload will go through Open vSwitch. DPDK is used to accelerate Open vSwitch, which performs better than the Open vSwitch (OVS) kernel data path and saves the CPU cycles for cloud computing server.

Linux kernel protocol stack provides the rich and powerful network service. However, it is also slow. The user space stack, on the top of DPDK, is highly desired for more performance boost. The attempt has been known, like applying BSD protocol stack in the user space, built on DPDK. Tencent, one of the biggest cloud computing companies, released this project, F-Stack, in open source, and it is available at http://www.f-stack.org/. Furthermore, the web application software like *Nginx*, and the memory database software like *Redis*, can be easily integrated with F-Stack. Nginx and Redis are very popular computing workloads running in the cloud data center.

1.5.3 ACCELERATED STORAGE

In 2016, Intel® open-sourced the storage performance development kit (www.SPDK.io). Its storage device is similar to NIC; it has I/O device, but for data storage, user space PMD is a more effective than Linux kernel driver for the high-performance scenario; the PMD can drive the NVMe device. It allows the application to have a faster access to the SSD. SDPK provides the other components such as NVMe-oF, iSCSI, and vhost server support. We will describe the storage and network acceleration in the later chapters.

1.6 OPTIMIZATION PRINCIPLES

While DPDK has adopted lots of optimization methods, many approaches are applicable for any software. The core principles are believed to be reusable in other area, which are as follows:

1. Target Software Optimization for a Specific Workload
   The specialized hardware is one way to achieve high performance, but DPDK uses the general-purpose processor and reaches the desired high-performance goal with software optimization techniques. In the early phase, the research efforts covered all possible platform components such as CPU, chipset, PCIe, and NIC. The optimization is always on the network workload characteristics.
2. Pursuing Scalable Performance
   The multicore era is a significant silicon progress. It enables high parallelism to achieve scalable performance. To avoid data contention, the system design needs to avoid the data race as much as possible. This is to say, design the data structure as a local variable, and use the lockless design to gain high throughout. Focus on an architectural approach to exploit the multicore for performance scaling.

3. Seeking Cache-Centric Design and Optimization

   In the context of system and algorithm optimization, code implementation optimization is much less known, often ignored. Code implementation optimization requires developers to have a good understanding of computer and processor architecture. DPDK depends on the optimization techniques such as cache usage and memory access latency impacts. As a programmer, if the code is written in a way to take cache utilization into account, the software optimization is probably half-finished. Most new software developers may not think how cache will behave at all.

4. Theoretical Analysis with Practice

   What is the performance limit? Is there any room for performance tuning? Is it worthy of in-depth study? Sometimes, it is easy to say, but difficult to do. By doing analysis, inferring, prototyping, and testing over and over, the optimization is often such an experimental journey. It is an incremental progress, and if the time permits, it is always good to have the performance model and analysis, as it will help to set achievable design goals.

Cloud computing is essentially taking more workload in a physical system. The success is due to CPU that has many core. Edge computing platform is the new trend, i.e., move the computing closer to the data, so it will provide the low latency computing experience, i.e., drive the new use case. DPDK can move data faster and store data quicker. DPDK, as the user space networking, is known for bypassing the heavy Linux kernel networking stack. Many of the DPDK optimization principles are also applicable to the Linux kernel network; there is an excellent progress in the kernel stack, such as XDP and AF-XDP. As Linux kernel comes up with its own bypass mechanism, more options are available for network developers to choose.

1.7   DPDK SAMPLES

DPDK concepts are discussed, and three examples are given here to get started with code for a quick look and feel.

1. *HelloWorld* is a simple example. It will set up a basic running environment for packet processing. DPDK establishes a software environment abstraction layer (EAL), which is based on Linux (alike) operating system, and causes this environment to be optimized for packet processing.
2. *Skeleton* is a most streamlined single-core packet sending and receiving example. It may be one of the fastest packets in/out testing code in the world.
3. *L3fwd*, Layer 3 forwarding, is one of main DPDK applications to showcase the use case, and it is heavily used for performance benchmark tests.

1.7.1   HELLOWORLD

HelloWorld is a simple sample for both codes and functions. It creates a basic running environment for multicore (multi-thread) packet processing. Each thread will print a message "hello from core #". Core # is managed by the operating system.

Unless otherwise indicated, the DPDK thread in this book is associated with a hardware thread. The hardware thread can be a logical core (*lcore*) or a physical core. One physical core can become two logical cores if hyper-threading is turned on. Hyper-threading is an Intel® processor feature that can be turned on or off via BIOS/UEFI.

In the code example, rte refers to the runtime environment and eal means environment abstraction layer (EAL). The most DPDK APIs are prefixed with rte. Similar to most of the parallel systems, DPDK has adopted the master thread and multiple slave threads models, which is often running in an endless loop.

```
Int
main(int argc, char **argv)
{
        int ret;
        unsigned lcore_id;

        ret = rte_eal_init(argc, argv);
        if (ret < 0)
                rte_panic("Cannot init EAL\n");

        /* call lcore_hello() on every slave lcore */
                RTE_LCORE_FOREACH_SLAVE(lcore_id) {
                rte_eal_remote_launch(lcore_hello, NULL,
                lcore_id);
        }

        /* call it on master lcore too */
        lcore_hello(NULL);

        rte_eal_mp_wait_lcore();
        return 0;
}
```

1.7.1.1   Initialize the Runtime Environment

The main entry of the master thread is <main> function. It will invoke the *below* entry function to start the initialization.

```
int rte_eal_init(int argc, char **argv).
```

The entry function supports the command line input, which is a long string of command combinations. One of the most common parameters is "-c <core mask>"; the core mask assigns which CPU threads (cores) need to be assigned to run DPDK master and slave threads, with each bitmask representing a specific core. Using "cat /proc/cpuinfo" can inspect the CPU cores on the given platform. A select core needs to be careful on the dual-socket platform, as local core and remote core can bring different workload performance.

As said, *rte _ eal _ init* includes a list of complicated tasks, e.g., parsing the input command parameters, analyzing and configuring the DPDK, and setting up the runtime environment. It can be categorized as follows:

- Configuration initialization;
- Memory initialization;
- Memory pool initialization;
- Queue initialization;
- Alarm initialization;
- Interrupt initialization;
- PCI initialization;
- Timer initialization;
- Memory detection and NUMA awareness;
- Plug-in initialization;
- Master thread initialization;
- Polling device initialization;
- Establishing master-slave thread channels;
- Setting the slave thread to the wait mode;
- Probing and initializing PCIe devices.

For further details, it is recommended to read the online document or even source code. One place to start is \lib\librte _ eal\common\eal _ common _ options.c. For DPDK users, the initialization has been grouped with EAL interface. The deep dive is only needed for in-depth DPDK customization.

1.7.1.2 Multicore Initialization

DPDK always tries to run with multiple cores for high parallelism. The software program is designed to occupy the logical core (lcore) exclusively. The *main* function is responsible for creating a multicore operation environment. As its name suggests, RTE _ LCORE _ FOREACH _ SLAVE (lcore _ id) iterates all usable lcores designated by EAL and then enables a designated thread on each lcore through rte _ eal _ remote _ launch.

```
int rte_eal_remote_launch(int (*f)(void *), void *arg,
unsigned slave_id);
"f" is the entry function that slave thread will execute.
"arg" is the input parameter, which is passed to the slave
thread.
"slave_id" is the designated logical core to run as slave
thread.
```

For example, int rte _ eal _ remote _ launch(lcore _ hello, NULL, lcore _ id). The parameter lcore _ id designates a specific core to run as a slave thread, and executes from the entry function lcore _ hello. In this simple example, lcore _ hello just reads its own logical core number (lcore_id) and prints out "hello from core #".

```
static int
lcore_hello(__attribute__((unused)) void *arg)
{
        unsigned lcore_id;
        lcore_id = rte_lcore_id();
        printf("hello from core %u\n", lcore_id);
        return 0;
}
```

In this simple example, the slave thread finishes its assigned work and quits immediately. As a result, the core is released. In most other DPDK-based samples, the slave
thread will run as an infinite loop, taking care of the packet processing.

1.7.2 SKELETON

This sample only uses a single core. It is probably the only DPDK sample to run with
a single core, and it is designed to implement the simplest and fastest packet in and
out of the platform. The received packets are transmitted out directly and without
any meaningful packet processing. The code is short, simple, and clean. This is a test
case to measure a single-core packet in/out performance on a given platform.

The pseudocode will call rte _ eal _ init to initialize the runtime environment, check the number of Ethernet network interfaces, and assign the memory
pool via rte _ pktmbuf _ pool _ create. Input parameter is designated with
rte _ socket _ id (), this is to specify which memory needs to be used, and
with no doubt, it always prefers the local memory in the local socket. We will explain
the basic concept later. Then, the sample calls port _ init(portid, mbuf _
pool) to initialize Ethernet port with memory configuration, and finally, it calls
lcore _ main() to start the packet processing.

```
int main(int argc, char *argv[])
{
        struct rte_mempool *mbuf_pool;
        unsigned nb_ports;
        uint8_t portid;

        /* Initialize the Environment Abstraction Layer (EAL). */
        int ret = rte_eal_init(argc, argv);

        /* Check there is an even number of ports to send/
        receive on. */
        nb_ports = rte_eth_dev_count();
        if (nb_ports < 2 || (nb_ports & 1))
        rte_exit(EXIT_FAILURE, "Error: number of ports must be
even\n");

        /* Creates a new mempool in memory to hold the mbufs. */
        mbuf_pool = rte_pktmbuf_pool_create("MBUF_POOL", NUM_
        MBUFS * nb_ports,MBUF_CACHE_SIZE, 0, RTE_MBUF_DEFAULT_
        BUF_SIZE, rte_socket_id());
```

```
        /* Initialize all ports. */
        for (portid = 0; portid < nb_ports; portid++)
        if (port_init(portid, mbuf_pool) != 0)
            rte_exit(EXIT_FAILURE, "Cannot init port %"PRIu8
"\n",portid);

        /* Call lcore_main on the master core only. */
        lcore_main();
        return 0;
}
```

1.7.2.1   Ethernet Port Initialization

```
port_init(uint8_t port, struct rte_mempool *mbuf_pool)
```

This function will be responsible for Ethernet port configuration like the queue configuration, and in general, Ethernet port is configurable with multi-queue support. Each receive or transmit queue is assigned with memory buffers for packet in and out. Ethernet device will place the received packets into the assigned memory buffer (DMA), and buffer is part of the memory pool, which is assigned at the initialization phase, socket aware.

It is important to configure the number of queues for the designated Ethernet port. Usually, each port contains many queues. For simplicity, this example only specifies a single queue. For packet receiving and transmitting, port, queue, and memory buffer configuration are the separate concept. If no specific configuration is specified, the default configuration will be applied.

Ethernet device configuration: Set the number of receiving and transmitting queues on a specified port, and configure the ports with input options.

```
int rte_eth_dev_configure(uint8_t port_id, uint16_t nb_rx_q,
uint16_t nb_tx_q, const struct rte_eth_conf *dev_conf)
```

Ethernet port/queue setup: Configure the specific queue of a specified port with memory buffer, the number of descriptors, etc.

```
int rte_eth_rx_queue_setup(uint8_t port_id, uint16_t
rx_queue_id,
                    uint16_t nb_rx_desc, unsigned int
socket_id,
                    const struct rte_eth_rxconf *rx_conf,
                    struct rte_mempool *mp)
int rte_eth_tx_queue_setup(uint8_t port_id, uint16_t
tx_queue_id,
                    uint16_t nb_tx_desc, unsigned int
                    socket_id,
                    const struct rte_eth_txconf *tx_conf)
```

After the Ethernet port initialization is completed, the device can be started with

```
int rte_eth_dev_start(uint8_t port_id).
```

Upon finishing, Ethernet port can have the physical MAC address, and the port will be turned on with promiscuous mode. In this mode, the incoming Ethernet packets can be received into the memory, allowing the core to do further processing.

```
static inline int
port_init(uint8_t port, struct rte_mempool *mbuf_pool)
{
        struct rte_eth_conf port_conf = port_conf_default;
        const uint16_t rx_rings = 1, tx_rings = 1;

        /* Configure the Ethernet device. */
        retval = rte_eth_dev_configure(port, rx_rings, tx_
rings, &port_conf);

        /* Allocate and set up 1 RX queue per Ethernet port. */
        for (q = 0; q < rx_rings; q++) {
                retval = rte_eth_rx_queue_setup(port, q,
RX_RING_SIZE,
                        rte_eth_dev_socket_id(port), NULL,
mbuf_pool);
        }

        /* Allocate and set up 1 TX queue per Ethernet port. */
        for (q = 0; q < tx_rings; q++) {
                retval = rte_eth_tx_queue_setup(port, q,
TX_RING_SIZE,
                        rte_eth_dev_socket_id(port), NULL);
        }

        /* Start the Ethernet port. */
        retval = rte_eth_dev_start(port);

        /* Display the port MAC address. */
        struct ether_addr addr;
        rte_eth_macaddr_get(port, &addr);

        /* Enable RX in promiscuous mode for the Ethernet
device. */
        rte_eth_promiscuous_enable(port);
        return 0;
}
```

The packet reception and transmission is done in an endless loop, which is implemented in the function of lcore _ main. It is designed with performance in mind and will validate the assigned CPU cores (lcore) and Ethernet devices that are physically on the same socket. It is highly recommended to use local CPU and local NIC in the local socket. It is known that remote socket will bring a negative performance impact (more details will be discussed in a later section). The packet processing is done with the packet burst functions. In both receive (rx) and transmit (tx) sides, four parameters are given, namely, ports, queues, packet buffer, and the number of burst packets.

Packet RX/TX Burst Function:

```
static inline uint16_t rte_eth_rx_burst(uint8_t port_id,
uint16_t queue_id, struct rte_mbuf **rx_pkts, const uint16_t
nb_pkts)

static inline uint16_t rte_eth_tx_burst(uint8_t port_id,
uint16_t queue_id, struct rte_mbuf **tx_pkts, uint16_t
nb_pkts)
```

Now we have the basic look and feel of DPDK packet receiving and transmitting code. The software has no dependency on vendor-specific NIC. From the very beginning, DPDK takes the software design into account, and the device abstraction layer is well designed to run across the platforms and NICs from multiple vendors.

```
static __attribute__((noreturn)) void lcore_main(void)
{
        const uint8_t nb_ports = rte_eth_dev_count();
        uint8_t port;
        for (port = 0; port < nb_ports; port++)
                if (rte_eth_dev_socket_id(port) > 0 &&
                        rte_eth_dev_socket_id(port) !=(int)
                        rte_socket_id())
                        printf("WARNING, port %u is on remote NUMA
                        node to "
                                "polling thread.\n\tPerformance
                                will "
                                "not be optimal.\n", port);

        /* Run until the application is quit or killed. */
        for (;;) {
                /*
                 * Receive packets on a port and forward them on
                the paired
                 * port. The mapping is 0 -> 1, 1 -> 0, 2 -> 3, 3
                -> 2, etc.
                 */
                for (port = 0; port < nb_ports; port++) {

                /* Get burst of RX packets, from first port of
                pair. */
                struct rte_mbuf *bufs[BURST_SIZE];
                const uint16_t nb_rx = rte_eth_rx_burst(port, 0,
                        bufs, BURST_SIZE);

                if (unlikely(nb_rx == 0))
                        continue;

                /* Send burst of TX packets, to second port of
                pair. */
```

```
        const uint16_t nb_tx = rte_eth_tx_burst(port ^
        1, 0,
                              bufs, nb_rx);

        /* Free any unsent packets. */
        if (unlikely(nb_tx < nb_rx)) {
                uint16_t buf;
                for (buf = nb_tx; buf < nb_rx; buf++)
                        rte_pktmbuf_free(bufs[buf]);
                }
        }
    }
}
```

1.7.3   L3FWD

This is a famous and popular DPDK example, as it is frequently used to measure DPDK performance metrics. The typical test scenario, server, is installed with the high-speed Ethernet adapters and connected with PCIe slots. Ethernet ports of the server platform are connected with the external hardware packet generator (usually from IXIA or Spirent), and the *l3fwd* sample can demonstrate 200 Gbps forwarding rate on dual-socket server platforms easily. In this sample, the packet is received from Ethernet. CPU will check the IP header for validation, and it will complete the routing table lookup using the destination IP address. Once the destination port is found, the packet is sent out with IP header modification like TTL update. Two routing table lookup mechanisms are implemented: the exact match based on the destination IP address and the LPM-based lookup. The *l3fwd* sample contains more than 2,700 lines of code (including blank lines and comment lines), and the main body is actually a combination of the HelloWorld and Skeleton.

To enable this instance, the command parameters are given in the following format:

```
./build/l3fwd [EAL options] -- -p PORTMASK [-P] --config(port,
queue, lcore)[, (port, queue, lcore)]
```

The command parameter is divided by "--" into two parts.

- The section after "--" is the command option of l3fwd sample.
- The section before "--" is used for DPDK's EAL options, mainly for run-time environment initialization and configuration.

[*EAL options*] is the configuration to set up the runtime environment, and it is passed over to rte _ eal _ init for processing.

- PORTMASK identifies the Ethernet ports for DPDK use. By default, the Ethernet device is managed by the Linux kernel driver. For example, the device name is "eth1". In today's DPDK version, the user can bind the specific device to DPDK, which uses the igb _ uio kernel module to allow the device configuration in the user space, where DPDK can take the device into control. A script is available to help the device bind operation, known as dpdk-devbind.py. The below example binds "eth1" for DPDK use.

```
dpdk-devbind --bind=igb_uio eth1
```

Note: In the early DPDK version, the DPDK initialization will scan the known PCIe devices for use, which can lead to the in-use network port being disconnected.

l3fwd sample configuration options a scalable performant approach on the basis of the (port, queue, lcore) configuration. It connects the assigned core with Ethernet port and queues. In order to achieve a high packet forwarding rate, multiple CPU cores can work together, and each core drives the specific port and queue for packet I/O (Table 1.2).

The master thread is similar to HelloWorld or Skeleton, and no further explanation is required here.

```
Initialize the running environment: rte_eal_init(argc, argv);
Parse the input parameters: parse_args(argc, argv)
Initialize lcore and port configuration
Initialize Ethernet ports and queues, similar to Skeleton
sample
Start the Ethernet port
Invoke the slave threads to execute main_loop ()
```

The slave thread will do the actual packet I/O, and the entry function is known as main _ loop(). It will run as follows:

```
Reads lcore information to complete configuration;
Reads information about send and receive queues;
Packet loop processing:
{
Sends packets in bulk to the transmit queue;
Receives packets in bulk from the receive queue;
Forwards packets in bulk;
}
```

Sending packets in bulk (or in bursts) to the designated queue and receiving packets in bulk from the designated queue are common in DPDK. It is an effective way for

TABLE 1.2

L3fwd Common Options: Port, Queue, Core

Port	Queue	Thread	Characterization
0	0	0	Queue 0 of processing port 0, thread 0
0	1	2	Queue 1 of processing port 0, thread 2
1	0	1	Queue 0 of processing port 1, thread 1
1	1	3	Queue 1 of processing port 1, thread 3

optimal platform resource use. Batched packet forwarding is done based on either the exact match (Hash) or the LPM selected as the compilation option. The example includes the code implementation based on SSE, known as "multi-buffer" principle, and is a known practice to get more performance on Intel® processors.

So far, lots of code have been shown here, and the intent is to give a quick feel on what's DPDK. In the later chapters, this book will not try much on the code deep dive. It is difficult to keep up to date with the latest code development, as DPDK community is still very active, so the used code might be obsolete as this book is in public.

1.8   CONCLUSION

What is DPDK? It is a set of software libraries, implemented on the basis of the software optimization principles and practices, hosted as an open-source project under Linux foundation. It is known for moving packets into server platform, and for moving packets into virtualization machines (or container tenants). The DPDK has been included by many Linux distribution packages, such as Red Hat, CentOS, and Ubuntu. DPDK is established as the leading user space networking open-source project. DPDK is widely adopted for networking and cloud infrastructure, network appliances, and virtual switch and storage acceleration systems worldwide.

FURTHER READING

1. https://wiki.linuxfoundation.org/networking/napi.
2. http://info.iet.unipi.it/~luigi/netmap/.
3. https://fd.io/2017/07/fdio-doubles-packet-throughput-performance-terabit-levels/.

2 Cache and Memory

Chen Jing
Alibaba

Heqing Zhu
Intel®

CONTENTS

DPDK is initially optimized for Intel®-based network platform. This chapter focuses on the cache, memory using Intel® processor based server platform as the hardware example. This chapter does not incorporate the non-Intel® platform; however, the concept should be very similar and applicable to other architecture and platforms.

2.1 DATA ACCESS AND LATENCY

Generally, the computer system consists of cache, memory, and storage hardware modules for data process, movement, and storage. Cache and DRAM hold the run-time data that the CPU needs to access, and the data will go away if the system is power off. Hard drives, SSD, optical disk, and USB flash drive are persistent data storage devices; once the data is written, it will exist in these devices even after the system is powered off. Cache and memory controllers have been an integrated part of modern processors. Typically, processors always access data from storage device to memory, then cache, and then finally load into CPU registers, and vice versa. In the networking case, if the data (packet) comes from Ethernet wire, NIC will do the DMA which places data into memory and cache, and data will finally be available in CPU registers for computation.

The processor needs to wait up to hundreds of clock cycles in order to read data from the memory (DRAM usually), and before data is available, the processor has to wait, and the waited cycles are known as the access latency. Usually, it is measured as the number of CPU cycles, the access latency is different, it is subject to where the data actually resides, and it also depends on the actual memory chip and interface. DRAM, SRAM, and Optane memory are the physical memory chips. They have the different access latency. Cache, usually SRAM, was designed for low access latency. L1/L2/L3 cache has a different latency. CPU loads data from the cache to the internal registers for the actual processing. The access latency reduction is important in developing a high-performance system, regardless of its computing, storage, or networking purpose. This chapter focuses on latency, with an in-depth discussion of cache, memory, and I/O.

When I started my engineering career, my work was related to 3G wireless system. At that time, telecommunication systems were known as the embedded systems: The platforms were always designed with limited memory (less than 25 MB) and a low-power CPU with low processing capability (such as the core that is running at 800 MHz). Real time is extremely important since the CPU speed is slow. The entire system like 3G wireless packet core gateway consists of 12 boards; each board is a separate subsystem that serves for specific and different functions, connected with a backplane using Advanced Telecommunications Computing Architecture (ATCA) chassis. Some boards have no general-purpose processors; the network processor or ASIC (application-specific integrated circuit) is the main logic to provide the fixed function. An example is a GTP-U data plane, and its IPsec function was implemented in a separate daughter card, which was used as needed, a plug-in option.

Nowadays, we actually see the communication system evolving at different paces. Some existing Telecom Equipment Manufacturers (TEM) are still selling the embedded systems on the basis of the legacy product architecture, but they are also working on the new product architecture. The other path is taken by cloud service provider; the software developers have built the similar network service with open source and servers, such as router, load balancer, and security system on the top of DPDK. The recent effort is to develop SD-WAN for WAN access optimization. Telecom operators (like AT&T) are deploying the networking functions on the server system, which is build out the telecom cloud to refresh the network infrastructure.

The underlying hardware change affects the software architecture. In the server system, the CPU is more powerful, cache and memory are much more abundant, and the platform system resource partition is done by virtualization or containerization technology. So the subsystem can run within a virtual machine. Software programmers need to make the following changes:

- Adapt the legacy telecom software (designed for the embedded system) to the server.
- Design new software architecture for telecom systems.

The legacy software is an important asset, but it is designed on the embedded system. It is not necessarily an optimized design/implementation when migrated to the server platform. In fact, the software optimization is desired to be done for the server platform.

2.2 INTEL® XEON ARCHITECTURE

Intel® launched the Xeon Scalable Processor (SP) in 2017. The product code name is Skylake, which is designed for data center workload. The high-end Xeon SP can have up to 28 physical cores within each CPU, and the low-end processor can have six physical cores at a minimum. Figure 2.1 describes what's inside of each physical core: It comes with L1 and L2 cache and memory channels. L1 cache has both data and instruction space with 32 KB each, and L2 cache size is about 1 MB.

Visiting the Intel® website can provide more information [1].

Figure 2.2 shows the multiple cores that are placed into a CPU package in generations. Each core has its own dedicated L1 and L2 caches. L3, as the last level cache (LLC), is a shared resource with all cores within the same CPU. Skylake architecture adjusts the size of L2 and L3 caches as system enhancement.

When core accesses the data in cache, the latency is different depending on the cache location (Table 2.1).

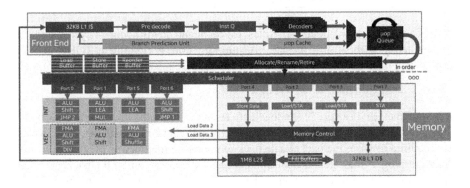

FIGURE 2.1   Intel® Xeon Processor core microarchitecture.

FIGURE 2.2   Intel® Xeon Processor cache architecture evolution.

TABLE 2.1
Intel® Xeon Processor Cache Access Latency

Cache Level	Category	Broadwell Microarchitecture	Skylake Server Microarchitecture
L1 data cache unit (DCU)	Size (KB)	32	32
	Latency (cycles)	4–6	4–6
	Maximum bandwidth (bytes/cycles)	96	192
	Sustained bandwidth (bytes/cycles)	93	133
	Associativity (ways)	8	8
L2 (mid-level cache (MLC))	Size (KB)	256	1024 (1MB)
	Latency (cycles)	12	14
	Maximum bandwidth (bytes/cycles)	32	64
	Sustained bandwidth (bytes/cycles)	25	52
	Associativity (ways)	8	16
L3 last level cache (LLC)	Size (KB)	Up to 2.5 per core	Up to 1.375[1] per core
	Latency (cycles)	50–60	50–70
	Maximum bandwidth (bytes/cycles)	16	16
	Sustained bandwidth (bytes/cycles)	14	15

L1 cache latency is about 4~6 cycles, L2 cache latency is about 14 cycles, and L3 cache latency is about 50–70 cycles. Assuming a core is running at 2 GHz, to load packet data from L3 cache, how many packets can be loaded per second?

```
2GHz / ~55 Cycles = 36.4 mpps
```

It is simple math: If a core has to load data from the L3 cache, then the limit is up to 36.4 mpps. If the core needs to do more tasks like packet header parsing, more cycles are required to write the packet data back to the cache/memory or I/O device. If the software programmer gets a clear and complete definition on all the packet-related operations, the overall processing cycles can be estimated in a rough manner. If the packet (data) is not present in the cache, the data will have to be loaded from the memory. But if the data is not present in the memory, then it has to be loaded from I/O device (via PCIe bus) or chipset. The memory access latency is significantly higher than the cache. And it has different access latency for a single-socket server or a dual-socket server. The dual-socket server will bring the NUMA (non-uniform memory access) concept.

Intel® Xeon Scalable Processor (shown in Figure 2.3).integrated two memory controllers inside the CPU, and each memory channel can be populated with three memory channels. DDR4 can support the frequency up to 2666 MHz. In a dual-socket system, two CPU processors are installed, allowing the server to be populated with lots of memory, and each socket supports up to 1.5 TB memory. It has a huge capacity, is flexible, and may be expensive. We have explained the concept of latency

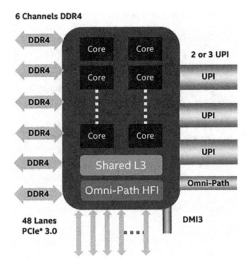

FIGURE 2.3   Intel® Xeon Scalable Processor architecture.

impact while using different memory devices; let's look further. The commonly used terms for memory are as follows:

- *RAM*: random access memory;
- *SRAM*: static RAM;
- *DRAM*: dynamic RAM;
- *SDRAM*: synchronous dynamic random access memory;
- *DDR*: double data rate SDRAM;
- *DDR2*: the second generation of DDR;
- *DDR3*: the third generation of DDR;
- *DDR4*: the fourth generation of DDR. This is supported by Intel® Xeon SP;
- *DDR5*: the fifth generation of DDR. Product is expected in 2020.

SRAM has much less latency than DRAM; it is more expensive, and has a small capacity and low power consumption; and it is heavily integrated within other chips such as CPU, FPGA (field-programmable gate array), and NPUs. Cache is usually made by SRAM, and it is important for latency-sensitive workloads. However, there is no flexible way to add more SRAM to the server platform. Different Xeon processors provide different cache sizes, but each comes with a different price. For example, for the second generation of Intel® Xeon SP (released in 2019), the cache size varies from 8 MB to 77 MB, so it is important to choose the right CPU. Cache latency is processor specific, and it is recommended to check the data sheet for details.

DRAM is usually the main system memory in server, DDR4 is the current mainstream DRAM technology, the DDR4 memory controllers are integrated into the processor, and the clock is running up to 2600 MHz. DDR5 RAM is expected to arrive in 2020. In the context of the dual-socket server (a common choice in the large-scale data center), NUMA is an important concept. As shown in Figure 2.4, each processor

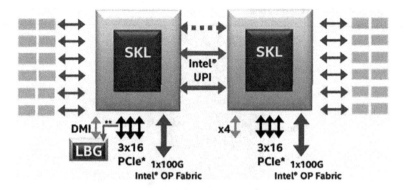

FIGURE 2.4   Intel® Xeon Scalable Processor—dual socket and NUMA.

(CPU) has the local memory controller with six channels, and it is obvious that data in the local memory has less latency (local socket) than data that resides in the remote memory (connected by the other processor, remote socket). The remote memory access requires accessing via a different bus (e.g., Intel® UPI). In the more complex system such as 4-socket system and 8-socket system, the remote memory access adds more latency. It is preferred to use the local memory for performance reasons.

In a deployment scenario, if the performance is given a high priority, the work-load placement needs to consider the NUMA awareness/optimization on the dual-socket server platform. In a real world, the server data center may have the mixed use of single-socket or dual-socket servers; if both platforms are deployed already, it requires the platform discovery mechanism (which can include NUMA awareness) to send the individual platform report to the cloud and NFV (network function virtualization) orchestration software packages such as OpenStack and Kubernetes. This solution is available in the market. Lots of software packages have the NUMA optimization. DPDK and Open vSwitch have the dedicated design to achieve NUMA optimization. How will NUMA be addressed in Kubernetes and the cloud-native approach? It is out of the scope of this book, and the latest Red Hat OpenShift may provide the commercial solution like k8s Node Manager.

2.3   INTEL® ATOM® SoC

Intel® launched Atom C3xxx SoC for networking and storage system in 2017. The code name is known as Denverton; the network series (NS) contains the network accelerator such as Intel® QuickAssist technology (QAT) and 4 × 10 Gbe integrated Ethernet within the chip. Visiting the website (http://ark.intel.com) will allow you to see all the processor choices. This processor line is good for the power-efficient system, and core counts are available from 2 to 16. Xeon processor is designed with out-of-order execution architecture, and atom is designed with in-order architecture. CPU frequency is available from 1.6 GHz to 2.6 GHz. AES-NI instruction is available for crypto computation.

For Intel® Atom Processor (C3xxx) shown in (Figure 2.5), only L1 and L2 caches are existing. L2 cache is shared by the core pair, and the cache size is up to the

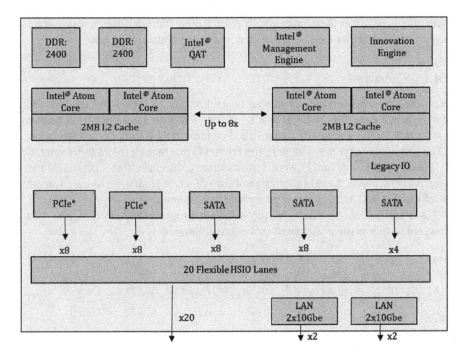

FIGURE 2.5   Intel® Atom C3xxx SoC.

different processors with 1 MB or 2 MB options being available. Hyper-threading is not supported. In the DPDK context, the physical threads are the same as the logical threads. Intel® Atom SoC is integrated with two DDR4 memory channels. The integrated Ethernet IP is based on 82599 and enabled by ixgbe PMD (poll mode driver) in DPDK. Roughly speaking, DPDK I/O workload running on atom core is about 50% performance of running a similar workload on Xeon/Xeon-D core. QAT, as crypto accelerator, can deliver up to 20 Gbps crypto/compression in this SoC; software library is supported by DPDK.

2.4   CACHE

The cache is introduced as the data latency reduction mechanism between processor and memory, and it enhances the data throughput significantly, as the program tends to access the local data more often. For a software programmer to make the best of cache mechanism, it is important to understand the concepts such as the *cache line, data prefetch, cache coherency, and its challenge* for software optimization. Most software programming books do not describe the software optimization much. The programming language is generally designed in an abstract way, independent of the underlying hardware platform. In a DPDK's software journey, the software was designed with the underlying platform awareness. In general, the software compiler has lots of command options to help the performance optimization; it requires the extra attention to use them properly. In general, the compiler does not provide the suggestion on whether the code is written in a cache optimization way.

2.4.1 CACHE LINE

In an Intel® processor, currently, there is no explicit way to allow software developers to manipulate data stays in L1, L2, or L3 cache. When data is loaded from memory, it is loaded in the chunks of cache line. For most ARM and x86 architectures, a cache line is about 64 bytes long. In the PowerPC machine, it is 128 bytes long. For the earlier ARM, x86, and PowerPC architectures, the cache line size was 32 bytes long. A cache line is the minimum size of data block of load/write to the cache; this is defined by CPU and is not programmable by software.

Knowing this cache line behavior, software programmer can design the data structure to follow the principle of cache line alignment.

```
#define RTE_CACHE_LINE_SIZE 64
#define __rte_cache_aligned
__attribute__((__aligned__(RTE_CACHE_LINE_SIZE)))

struct rte_ring_debug_stats {
        uint64_t enq_success_bulk;
        uint64_t enq_success_objs;
        uint64_t enq_quota_bulk;
        uint64_t enq_quota_objs;
        uint64_t enq_fail_bulk;
        uint64_t enq_fail_objs;
        uint64_t deq_success_bulk;
        uint64_t deq_success_objs;
        uint64_t deq_fail_bulk;
        uint64_t deq_fail_objs;
} __rte_cache_aligned;
```

In fact, the compiler is intelligent. It will attempt to align each data structure with a cache line as part of compiling optimization. If legacy software code was initially designed (or tuned) for a specific processor architecture, the code may contain the programming directive to discourage such optimization. There is a reason for this: In earlier days, the memory was a limited resource in the embedded system, and the memory-optimized design was an important system constraint, which matters for those early programmers.

2.4.2 CACHE PREFETCHING

Cache prefetch is a predictive action. It speculates to load the data that will be used soon. "C" programming language has not defined an explicit function to do the data prefetch. You may have seen that the prefetch option is quite common in the system BIOS, which can turn on/off the CPU prefetcher. Most CPU will speculate what's the next data/instruction to be used, the speculative mechanism will load the instructions/data from memory to cache, and it can be done without software code to make a request. But it can also be done with software's explicit ask. As such, data locality is one design principle; if the instructions or data may be temporarily stored in the cache, there is a higher chance to be used by CPU again. A software loop is such an example: CPU will repeatedly execute the instructions until loop exits. Spatial locality is another example; it refers to the instruction or data to be used that may be adjacent to the instruction or data being used now. As said, a cache line is designed as 64 bytes long. By loading more data from the adjacent memory, it can reduce the access latency for future processing; hence, it helps the performance.

However, there is a risk due to the nature of speculative execution. If the prefetch data is not the actual data that the CPU will use, the prefetch can pollute the limited cache space. Another load will be issued later, and in the worst case, this may lead to performance degradation. In some cases, even if the hardware prefetcher is activated, the program may not see the benefits. It is workload specific, and the result depends on the data and code execution. Likely you may (or not) see some system optimization guides to suggest: Turn off prefetch as part of the system tuning practice.

Packet processing is such a workload that gains a performance boost by using the prefetch. This is visible in many DPDK code samples like l3fwd. Prefetching is interesting topic; let's look at the code sample to get a hint on how to write an efficient code. The two code samples defined assign values to a two-dimensional array: arr[1024][1024]; the software implementation is different for the inner loop.

```
Sample-1
// Assigns value to a[i][0], a[i][1], a[i][2] ... a[i][1023]
for(int i = 0; i < 1024; i++) {
        for(int j = 0; j < 1024; j++) {
            arr[i][j] = num++;
        }
    }
Sample-2
// Assigns value to a[0][i], a[1][i], a[2][i] ... a[1023][I].
for(int i = 0; i < 1024; i++) {
        for(int j = 0; j < 1024; j++) {
            arr[j][i] = num++;
            }
}
```

For the Sample-1, the code logic is closer to make use of the cache line, as the next data access is within the cache line; it can reduce memory access latency because of the prefetch. For code Sample-2, the array access poses a challenge as the next data is far beyond the next cache line. This program may experience more memory latency, which can lead to low execution efficiency, and CPU stall when waiting data to be load into the cache (Figure 2.6).

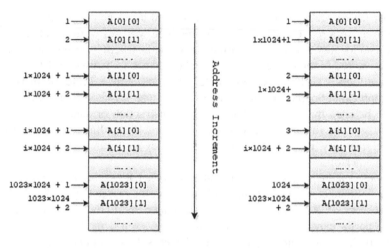

Data Access in Sample-1   Data Access in Sample-2

FIGURE 2.6   Data prefetch code samples.

2.4.3 SOFTWARE PREFETCHING

x86 CPU allows the software to give hints for the data prefetcher. As said earlier, there is no known standard "C" language API. This feature may not be known to all the software programmers. This tip is used in DPDK to develop the performance critical software (Table 2.2).

Increasing the use of prefetching is not the same as controlling the data to stay in the cache; it allows intentionally loading the data to the cache, which saves core wait as it is in the idle mode. However, the incorrect use of the prefetching may increase the unwanted data in the cache and decrease the performance. More seriously, it may affect other programs running in the system as L3 cache is a shared resource for all cores. If a large amount of data is loaded into L3 cache by one core, and if L3 cache is filled at some level, future data loaded will evict the current data out of cache. The further processing on the evicted data is going to experience the extra latency, as it will take hundreds of cycles to load the evicted data back. Therefore,

TABLE 2.2

Prefetch Instruction Details

Instruction	Descriptions
PREFETCH0	Prefetch data into all levels of the cache hierarchy.
PREFETCH1	Prefetch data into level 1 cache and higher.
PREFETCH2	Prefetch data into level 2 cache and higher.
PREFETCHNTA	Prefetch data into nontemporal cache structure and a location close to the processor, minimizing cache pollution.

TABLE 2.3
Memory Prefetch Intrinsic

Instruction	Prefetching Instruction
_MM_HINT_T0	PREFETCH0
_MM_HINT_T1	PREFETCH1
_MM_HINT_T2	PREFETCH2
_MM_HINT_NTA	PREFETCHNTA

software developers should design and test carefully before releasing code into the final product (Table 2.3).

Prefetching instructions are the assembly instructions. For many software developers, it is inconvenient to insert the assembly code. The extra software libraries provide such APIs. For example, "mmintrin.h" is given below:

```
void _mm_prefetch(char *p, int i);
p: the memory address that needs to be prefetched;
i: the corresponding prefetching instruction;
```

In DPDK, PMD has used prefetching instructions heavily. DPDK can get one core to handle up to 33 mpps, on average, and it is about 80 cycles to process a packet. Let's have a look to see what has been done.

1. Write NIC RX descriptors into memory, and fill the descriptor with the data buffer pointer. Upon receiving a packet, the NIC will write the packet into the specified buffer.
2. Read the RX descriptors in the memory to check if a packet is received (the NIC will update the descriptors as soon as it receives the packet, memory read). After the packet arrival is confirmed, read the pointer of the control structure from the memory, and then read the control structure itself again from the memory (memory read). Fill the RX descriptors information to the control structure.
3. Update the NIC RX queue (RXQ) to indicate that a new packet is ready to be processed.
4. Read the packet header (memory read), and decide the next port for Packet TX.
5. Fill the packet information to the control structure in the NIC TX queue (TXQ) descriptors. Update the TXQ register.
6. Read the NIC TX descriptor from the memory (memory read), and check whether there are packets waiting to be transmitted by hardware.
7. If there are packets waiting, read the control structure from memory (memory read) and release the data buffer.

There are six times of "memory read" for a packet handling procedure as shown above. As discussed earlier, data location matters. CPU needs 3–5 clock cycles to read data from the L1 cache, more than 12 clock cycles from the L2 cache, ~50+ clock cycles from the L3 cache, and hundreds of clock cycles from the memory. There are only 80 cycles as the total budget to process a packet. It does not add up, right? If the data is already loaded in the cache, memory reference can be translated to the cache access, and high latency becomes low latency. The prefetching is the secret here, so it is extremely useful for software driver optimization. The other main helper comes from Intel® DDIO (Data Direct I/O), which enables the NIC to receive the packet into cache directly. Without DDIO, NIC receives the packet from the NIC to the assigned host memory, and then the CPU loads the packet from the memory to the cache, eventually using the packet data via CPU's own registers.

The memory reading on NIC control structures and packet data buffer are not guaranteed by CPU-based prefetching. The software driver can enhance it with software hints, prefetching the data in advance. Here is the code sample with highlighted prefetch actions:

```
while (nb_rx < nb_pkts) {
rxdp = &rx_ring[rx_id]; //read RX descriptor
staterr = rxdp->wb.upper.status_error;

//check if the packet is received or not
if (!(staterr & rte_cpu_to_le_32(IXGBE_RXDADV_STAT_DD)))
break;
rxd = *rxdp;
//allocate data buffer
nmb = rte_rxmbuf_alloc(rxq->mb_pool);
            nb_hold++;
//read controlling structure
            rxe = &sw_ring[rx_id];

            ......

            rx_id++;
            if (rx_id == rxq->nb_rx_desc)
            rx_id = 0;

//prefetch the next ctrl structure's mbuf
rte_ixgbe_prefetch(sw_ring[rx_id].mbuf);
// Prefetch RX descriptor and ctrl structure ptr
if ((rx_id & 0x3) == 0) {
                rte_ixgbe_prefetch(&rx_ring[rx_id]);
                rte_ixgbe_prefetch(&sw_ring[rx_id]);
            }
            ......
//prefetching packet
rte_packet_prefetch((char *)rxm->buf_addr +
rxm->data_off);

//store the RX queue info to the Ctrl structure mbuf
rxm->nb_segs = 1;
rxm->next = NULL;
rxm->pkt_len = pkt_len;
rxm->data_len = pkt_len;
```

```
rxm->port = rxq->port_id;
......
rx_pkts[nb_rx++] = rxm;
```

2.4.4 FALSE SHARING

The LLC (L3) has a larger capacity, and it is shared by all cores. In certain cases, the same memories are accessed by multiple cores, causing conflict to occur when multiple cores write or read the data *in the same cache line*. x86 is designed with a sophisticated mechanism to ensure cache coherency, and software programmers can enjoy such CPU features without worrying about data contention and corruption in a multicore running environment. There is a cost for the data contention in a cache line, and if multicore is trying to access the different data in the same cache, CPU will invalidate the cache line and force an update, hurting the performance. This data sharing is not necessary because the multiple cores are not trying to access the same data, which is known as false sharing. The compiler can find the false sharing, and it will try to eliminate the false sharing at the optimization phase. If compiler optimization is disabled, then there is no compiler attempt to work on this problem.

There is legacy code, it is initially designed for single-core system, and some global variables are used for statistical purposes. When the code is migrated to run in a multicore environment, multiple cores may update the global variables, leading to a "false sharing" issue. Code inspection can help to identify the false sharing, but sometimes it is difficult if the code base is huge and code reading is time-consuming. Intel® software tools, VTune, and/or Intel® PTU can help to identify the runtime false sharing issue.

If false sharing is identified, the software fix is simple: Just ensure the shared data elements reside in a different cache line. If a data structure is used for multicore use, each individual data member can occupy its own cache line. The below example is designed to support four cores to update the packets statistics, and it is designed to avoid false sharing.

```
#define RTE_CACHE_LINE_SIZE 64
#define __rte_cache_aligned
__attribute__((__aligned__(RTE_CACHE_LINE_SIZE)))

struct rte_port_core_stats {
        __rte_cache_aligned uint64_t core1_pkts;
        __rte_cache_aligned uint64_t core2_pkts;
        __rte_cache_aligned uint64_t core3_pkts;
        __rte_cache_aligned uint64_t core4_pkts;
};
```

2.4.5 CACHE COHERENCY

The cache coherence is handled by CPU since this reduces software complexity. There is a performance hit if multiple cores need work on the same cache line. DPDK is expected to be deployed on a multicore environment. One basic rule is to avoid multiple cores to access the same memory address or data structure as much as possible.

Example 1: Design the per-core-based data structure. Minimize the data access on its own core granularity.

```
struct lcore_conf {
      uint16_t n_rx_queue;
      struct lcore_rx_queue
rx_queue_list[MAX_RX_QUEUE_PER_LCORE];
      uint16_t tx_queue_id[RTE_MAX_ETHPORTS];
      struct mbuf_table tx_mbufs[RTE_MAX_ETHPORTS];
      lookup_struct_t * ipv4_lookup_struct;
      lookup_struct_t * ipv6_lookup_struct;
//cache line alignment
struct lcore_conf lcore[RTE_MAX_LCORE] __rte_cache_aligned;
```

The data variable of "struct lcore _ conf" is always aligned by the cache line, so that data will not cross two cache lines, reducing the risk of multicore access scenario. In the array of "lcore[RTE _ MAX _ LCORE]", all cores are guaranteed to data access in separate cache lines, avoiding the contention conflict.

Example 2 is the NIC-related design, NIC supports multiple queues, and each queue might be assigned by different cores for packet processing (Figure 2.7).

When multiple cores want to access the same NIC, DPDK usually will prepare one RXQ and one TXQ for each core to avoid the race condition. The below diagram

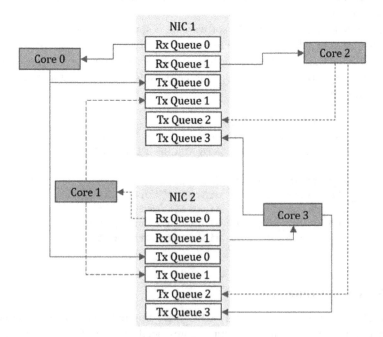

FIGURE 2.7 Multiple queues and multiple cores.

shows four cores that might access two NIC ports simultaneously. NIC 1 and NIC 2 both have two RXQs and four TXQs: cores 0, 1, 2, and 3. Each has a RXQ and a TXQ. Core 0 receives data from the RXQ 0 in the NIC 1 and can send it to the TXQ 0 in the NIC 1 or the TXQ 0 in the NIC 2. Similarly, core 3 receives the data from RXQ 1 in the NIC 2 and can send it to the TXQ 3 in NIC 1 and the TXQ 3 in NIC 2.

2.4.6 Noisy Tenant and RDT

When multiple cores are competing for the shared cache resource, some cores may repeatedly ask for more cache resources, and the remaining cores are forced with less cache to use; the aggressive core is somewhat acting as the role of the noisy neighbor. This is not a real problem if the whole system is utilized by single owner, who will make the decision to assign resources correctly. In public cloud computing, where resource is shared to accommodate multiple guests' workloads together, avoiding the noisy neighbor is important because all tenants are paid for the resource. It is highly desirable to provide a more granularity control on using the shared cache resource. Intel® proposed RDT (Resource Director Technology) framework to tackle the "noisy neighbor" problem, and RDT includes multiple technology ingredients such as cache monitoring, cache allocation, memory bandwidth monitoring, and cache and data optimization technologies. RDT is not specific to DPDK, so it is not introduced in this book.

2.5 TLB AND HUGEPAGE

In the computing system, software uses the virtual memory address, not the physical memory address. Memory management and paging technology has been widely used for address translation. System memory is organized in pages, and the traditional memory page size is 4 KB. HugePage is later introduced and Linux supports the hugepage size as 2 MB or 1 GB. The memory translation is indeed a multilevel page table lookup. TLB (translation lookaside buffer) is part of the CPU. It can speed up the memory address translation, and it is a cache for virtual memory to physical memory. On a given virtual address, if the entry resides in the TLB, the physical address will be found immediately on a given virtual address. Such a match is known as a TLB hit. However, if the address is not in the TLB (TLB miss), the CPU may do the page walk to finish the address translation, which takes long cycles as it may do the multiple memory references; if the page table is not available in cache, the miss will lead to the memory access, which depends on the access latency, potentially goes up to hundreds of cycles.

In the latest Skylake processor, TLB entries are supported for 1 GB page size and 32 entries are supported for 2 MB page size. 4 GB memory is sufficient for many networking workload uses. The DPDK practice is to set up a hugepage for packet processing memory use. Assign hugepage memories, and each is configured with 1 GB size at Linux configuration. The TLB hit will avoid (or eliminate) the need for page walk, thus avoiding extra latency.

Linux supports a file system known as hugetlbfs, which is designed to support the HugePages of 2M or 1G page size. In order to use HugePages, the hugetlbfs must be enabled while compiling the Linux kernel. If hugetlbfs is enabled, the early

DPDK software version will reserve a certain amount of memory as HugePages. There are two ways to reserve HugePage using Linux.

- Reserve the HugePage at Linux system initialization time and it will take this as the launch command option.
- Dynamic memory reservation after the Linux is launched.

2.5.1 RESERVE HUGEPAGE AT BOOT TIME

To reserve 4 GB memory as HugePages (each size is 1 GB), the following command can be used at Linux launch option.

```
default_hugepagesz=1G hugepagesz=1G hugepages=4
```

It should be noted that the system can support HugePage and the maximum supported size of HugePages depends on the actual processor support. Linux software programmer can read /proc/cpuinfo to see the actual feature availability.

- Bit PSE indicates 2 MB HugePage support.
- Bit PDPE1GB indicates 1 GB HugePage support.

Different processor architectures may support the different HugePage size. The huge page memory is a shared system resource. For NUMA-aware system, when reserving the HugePages in 1 GB page, one practice is to reserve 2 GB memory in NUMA node 0 and to reserve another 2 GB memory in NUMA node 1.

2.5.2 RESERVE HUGEPAGE AT RUNTIME

After Linux has been launched, the following Linux command can be used to reserve 1024 pages of 2 MB memory (2 MB HugePages).

```
echo 1024 > /sys/kernel/mm/hugepages/hugepages-2048kB/
nr_hugepages
```

In the NUMA system with two nodes, we can use the following command:

```
echo 1024 > /sys/devices/system/node/node0/hugepages/
hugepages-2048kB/nr_hugepages
echo 1024 > /sys/devices/system/node/node1/hugepages/
hugepages-2048kB/nr_hugepages
```

For some Linux systems that do not support 1 GB HugePage by default, the huge page reservation needs to be enabled at Linux boot time. After the HugePages have been reserved, they need to be mounted before use, like a path/mnt/huge.

```
mkdir /mnt/huge
mount -t hugetlbfs nodev /mnt/huge
```

It should be noted that before mounting, we must ensure that the memory is reserved already, or else the above command will not work. It will mount the file system temporary.

The other approach is to modify /etc/fstab by adding the line:

```
nodev /mnt/huge hugetlbfs defaults 0 0
```

For 1 GB HugePages, we must use the following command:

```
nodev /mnt/huge_1GB hugetlbfs pagesize=1GB 0 0
```

DPDK and HugePage—2018 Changes

Before 2018, DPDK memory system was designed on the basis of the hugepage. At the system initialization phase, mmap(), a system call, will map the reserved HugePages to the virtual address space in the user space. Then, the HugePages can be used by DPDK application in the user space application. Using this, DPDK application can reduce the penalty of TLB miss. It is good for performance-critical software.

But there are also drawbacks. HugePage reservation is good for performance, and it is a system privilege configuration that may not fit well with the cloud-native concept. In cloud-native principle, network service is designed for scale-up and scale-down easily. When the workload is high, more service instances can be spawned. When the workload is reduced, the service instance can be reduced, even completely disappearing. The service is expected to run within a container, and when applying DPDK in a container, without privilege, the container has no hugepage for use.

In 2018, DPDK18.05 introduced a new approach to reserve the HugePage at DPDK runtime. The trade-off is to live with 4 KB page memory. DPDK memory system is modified without rigid dependency on the hugetlbfs.

Many network applications have existed for many years. The legacy software applications have their own memory management system, even if they are less performant than the optimized memory design in DPDK, but it may have the rich functions such as memory diagnostics feature and high granularity of memory allocation algorithms. When those legacy applications want to take advantage of DPDK, one way is to avoid using the DPDK memory buffer and management interface. So the existing software can still migrate to DPDK-based design and integrate its own memory management module with DPDK memory management subsystem. The external memory module is considered as "external" memory management.

2.6   MEMORY LATENCY

In reality, cache size is a limited resource. In the cache miss scenario, the DRAM access latency cannot be hidden, and then it will go up to a few hundred cycles. It is different if the memory access is in the local or remote socket. Intel® memory latency checker (Intel® MLC) is a software utility to measure the memory latency and bandwidth. It can also be used to measure cache-to-cache latency. In the idle workload, if the local memory access takes 67 ns, then by multiplying it with core frequency (2.5 GHz), we can estimate as 67 * 2.5 = 167.5 cycles.

If the remote socket memory access is 126.5 ns, then by multiplying it with core frequency (2.5 GHz), we can estimate as `126.5 * 2.5 = 316.5 cycles`.

In the heavy workload system, the memory latency can deteriorate further up to 190 ns. The software utility is open for public download, and it is helpful to estimate the system performance and diagnose the memory metrics in the NUMA system [2].

2.7   DDIO

Intel® DDIO is available on Intel® Xeon Processor (Sandy Bridge) since 2012. In the networking context, the packet data arrives on the NIC (I/O device) first. Traditionally, I/O devices have the DMA capability to move packets into memory directly; when CPU needs to process the packet, it will be loaded from the memory to the cache, and this means the extra access latency. Intel® DDIO enables the NIC and CPU to exchange data directly through the L3 cache, and NIC may move the packet into the LLC (instead of memory), as the new primary destination. In an optimized system, it reduces about 100 cycles for the arrived Ethernet packets. The same is true at the transmit side. Intel® DDIO effectively reduces the memory access latency, the software programmer gets it free, and it does not require any software change. The DDIO will not move every incoming packet directly into the LLC, and it will depend on the available space in the LLC. If there is not enough space to accommodate the new packet, NIC will move packets to the memory.

2.8   NUMA

For the dual-socket server, there is a local and remote memory. Hence, NUMA matters, how the software (where it is running) will allocate/use the memory? Remote memory access will add the access latency and will hurt the performance.

- *Local core and local memory*: Per-core memory has been illustrated in the false sharing and cache coherence section. The per-core memory means that each core uses its local memory for the frequently used data structure on its own, avoiding the multicore to compete the data whenever possible.
- *Local device with local core*: The device is physically connected to a PCIe slot, which is subject to a local socket with many cores, and the principle is to use the local cores' local memory to process the packet data on the local PCIe device. If there is a PCI device on NUMA node 0, we should use the core(s) on this "NUMA" node for the device. The designed data structure and the assigned packet buffers are allocated from the local memory. The following is an example of allocating the local memory:

```
/* allocate a cache memory for the queue structure */
q = rte_zmalloc_socket("ixgbe", sizeof(*q), RTE_CACHE_
LINE_SIZE,  socket_id);
```

FURTHER READING

1. Intel® Xeon Scalable Processor SKU details at https://ark.intel.com/products/series/125191/Intel-Xeon-Scalable-Processors.
2. https://software.intel.com/en-us/articles/intelr-memory-latency-checker.

3 Core-Based Parallelism

Qun Wan, Heqing Zhu, and Zhihong Wang
Intel®

CONTENTS

Gain more performance without any software change? Prior to 2006, this was easily done with just a simple CPU upgrade. In those early years, new CPUs provide a higher frequency and an enhanced IPC (instructions per cycle) that would allow the software to run faster. Columbia University published the online course material, which summarized the performance increase on a known "SPECInt" benchmark since 1978, the very beginning of the processor era. It was about a 25% performance increase per year until 1986. From 1986 to 2002, it became about a 52% performance increase per year, largely driven by clock frequency and micro-architecture enhancement. From 2002 to 2006, it went back to a 20% performance increase per year. A simple trip to the Intel® Museum (Santa Clara, CA) would be fun not only to learn the processor history but also to see the actual historic chip sample [1].

About the early 2000, the silicon industry acknowledged hitting the wall of physics: Increasing the frequency on a tiny room is very challenging, as the linear frequency growth causes the steep power dissipation to increase, exceeding the power budget limit on a tiny chip space. The chip industry faced this significant challenge, and as a result, the processor jumped into a new era of multicore, and the multicore architecture became a silicon foundation, which paved the way for cloud computing, Software Defined Network(SDN)/ Network Function Virtualization(NFV), etc.

3.1 MULTICORE PERFORMANCE

3.1.1 HORIZONTAL SCALE

A multicore processor refers to a chip consisting of two or more cores (or computing engines). It is a vertical scaling approach performed by increasing the frequency to boost processor performance, and a horizontal approach to scaling-up performance is done by adding more cores. To take advantage of more cores, the software model needs to be adapted and designed in a multicore-friendly way. For the tasks executed on a single core in a sequential order, they won't get a performance boost as a new processor runs at the same frequency as the old one, but more cores. In order to get the performance boost, the software design needs to be revisited to take advantage of the multicore architecture. For example, how to logically divide a big task into several subtasks, then assigning the multiple subtasks on different cores, resulting in the total tasks done in parallel.

Simply speaking, if the software is not designed for multicore, its performance may not gain improvement from more cores running on the same clock frequency. If the software was released in binary, it does not allow the code to modify for multicore architecture. There was a short period when lots of application performances were not enhanced even the hardware system is upgraded with a multicore processor.

Software needed to gain the performance benefits from the increasing core count of the processor, and a fresh look is required. *Amdahl's law* is a serious study of this computer architecture model back in 1967. It tells us that the theoretical speedup in latency of the execution of a program is a function of the number of processors executing it. The speedup is limited by the serial part of the program. In other words, the latency of multicore parallel computing cannot be infinitely small with an increased number of cores. This law clearly indicates that the key to performance increase on a fixed workload needs to be focused on reducing the proportion of the sequential execution part in the whole task. For more information, see http://www.d.umn.edu/~tkwon/course/5315/HW/MultiprocessorLaws.pdf.

Furthermore, Gustafson's law points out that true parallel power of a large multiprocessor system is only achievable when a large parallel problem is applied. Packet processing, the main use case of DPDK (Data Plane Development Kit), throughput is a key performance metric. DPDK needs to deliver the packet forwarding throughput, which increases linearly with the increased number of cores used. The law guides the DPDK design principles and code implementations:

- The resource localization;
- Avoidance of cross-core sharing;
- Reduced data collision in the critical zone;
- Fast execution in the critical zone.

The last two items involve multicore synchronous mechanism and will be explained Chapter 4. The key idea is to reduce the proportion of nonparallel processing and concurrency interfered to varying degrees.

3.1.2 MULTICORE

In this section, we will take Intel® Xeon multicore processors as an example to visit the basic concepts used in DPDK, such as physical core, logical core, and CPU node.

The following diagram describes the single-core, multicore, and hyper-threading (HT) concepts. Figure 3.1 describes the single-core microarchitecture, and the core contains the instruction perfecter, decoder, and ALU execution units. The cache is part of core.

The multicore architecture is everywhere in most of the computing systems. Figure 3.2 gives a high-level view of x86-based architecture. Within the processor, each core may have the internal execution units and dedicated cache. Uncore includes the shared cache (also known as the last level cache, L3 cache), integrated

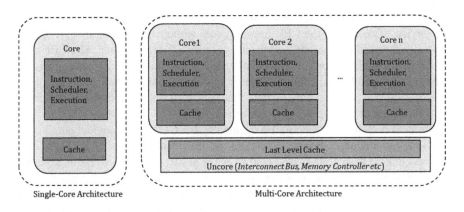

FIGURE 3.1 From single to multicores.

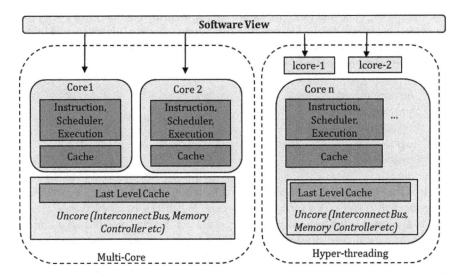

FIGURE 3.2 Software perspective on multicore vs hyper-threading.

memory controller, and interconnect bus for connecting multiple nodes, which can build two sockets, or even four or more sockets in server platform.

On multicore architecture in a dual-socket platform, a multi-threading software program can run on separate cores. The thread coordinates among cores and processors (nodes, sockets) are realized via the uncore unit (shared cache, interconnect bus, the local and remote memory), and the overhead arises from the cache coherency for shared data. From the memory access perspective, NUMA (non-uniform memory access) is now part of the system, which was introduced in early chapters, and the NUMA adds the latency if remote memory is used, causing the system performance variation. The `socketid` parameter, representing the physical processor (socket), is added in the software interface design. It is added to avoid the remote memory, and its parameter is implemented everywhere in the DPDK source code.

The CPU is running faster than the data access through external memory (e.g., I/O device). The core in the CPU may stay idle while waiting for data arrival. HT is designed to reduce the CPU idle states. Figure 3.2 explains HT. The CPU supports HT technology, and any physical core may contain logical cores 2 (which can be 4, 8, depends on the CPU). When HT is enabled, the software sees two logical cores on one physical core. The technology is to reuse the multiple execution units of the superscalar pipeline more effectively.

Physical core and HT support are decided by the CPU. Every logical core has its complete and independent register set, and includes local interrupt logic. There is no difference other than a physical thread from the software perspective. For example, HT technology will make an 8-core processor work like 16 logical cores that are available for software execution. Using the HT technology, the overall execution resource utilization can be enhanced; therefore, multi-thread software can have a higher performance gain.

However, if the software is written with high computation (high demand for IPC) tasks, which does not wait much for external data, there are not many idle cycles that the core needs to wait for data arrival, the CPU won't switch to idle much, and in this context, the HT will not bring significant performance benefits.

So far, it might be confused to see

- Software thread;
- Hardware thread;
- Logical thread;
- Logical core;
- Physical core.

When HT is supported and enabled (via BIOS setting), two (or more) logical cores are presented to software over a physical core. When HT is disabled, one logical core is presented to software over one physical core. Hardware thread is a similar term to the logical core. Software thread is the typical software concept, in Linux-alike operating system, and each software process can spawn multiple threads for high parallelism applications. It is the responsibility of the OS to schedule software threads to specific cores.

TABLE 3.1

Linux CPU Command Tools

CPU Information	Command
The number of cores (physical cores)	`cat /proc/cpuinfo	grep "cpu cores"	uniq`

The number of logical cores	`cat /proc/cpuinfo`
	`If "siblings" and "cpu cores" are consistent,`
	`it indicates that hyper-threading is not`
	`supported or disabled. If "siblings" is double`
	`to "cpu cores", it indicates that hyper-`
	`threading is supported and enabled.`

Socket ID	`cat /proc/cpuinfo	grep "physical id"| sort|` `uniq| wc -1 or lscpu	grep "CPU socket"`
Get ID for processor	`cat /proc/cpuinfo| grep "processor"| wc -1`

As a software developer, the above command can help to understand what's available at a Linux system. CPU core information is given in Table 3.1.

The number of physical cores: Processor cores, or CPU cores, is the number of cores in each physical CPU. For instance, the Intel® Xeon Platinum 8180 @ 3.80 GHz, launched in Q3'2017, is a 28-core processor with 28 processor cores inside. Logical cores with the same "core ID" are hyper-threads of the same core.

The number of logical cores (lcore_id): The SIBLING is the number of hyper-threads that are owned by a single physical processor from the Linux kernel, namely, the number of logical cores in one physical package. If the SIBLING equals the number of actual physical cores, it indicates that HT is not enabled, and vice versa.

Socket ID (socketid): The number of sockets is the number of physical processors (CPUs) in the server system. The below function provides the relationship between the socket and the logical core.

```
socketid = (uint8_t)rte_lcore_to_socket_id(lcore_id);
```

3.1.3 Affinity

As the processor has the multicore architecture, by default, the OS task scheduler is responsible for running software on the cores. OS has an advanced mechanism, such as supporting the processor affinity (also known as CPU pinning), which allows a particular task (process) to run on the given core(s). This will give software with the dedicated core(s) as the execution resource.

Without the affinity mechanism, a software thread can be moved from one core to another arbitrary one; this approach is effective for flexible use, such as resource balancing case, but is not optimal for performance critical use case. Moving software thread from one core to another is associated with the system overhead such as the negative impacts on using cache.

For performance-sensitive software like DPDK, the CPU affinity is being heavily used in its software APIs. For example, L3fwd, a DPDK sample, will explicitly specify the dedicated core(s) to for packet in/out processing on NIC Rx/Tx queues.

1. Linux Kernel's Support

    In the Linux kernel, there is an important data structure called `task_struct`, which relates to the software task (process) definition. `cpus_allowed` is a bit mask that corresponds to the logical cores (cpus) in the system. By default, in Linux, a process can run on any process and migrate among processors when necessary. The Linux provides the `taskset` and the below software APIs for CPU affinity assignments.
    - `sched_set_affinity()` (Assign the process with CPU affinity);
    - `sched_get_affinity()` (Get the CPU affinity on the specified process).

    Please note that `cpu_affinity` may be passed on to the child thread, so before using the `sched_set_affinity`, a careful design consideration is required.

2. Why Use Affinity?

    The most obvious benefit of binding the software task to the dedicated execution resource, which is indeed an optimal use of core resources if the performance is the key goal. Each core has its own dedicated L1/L2 cache. It keeps the most recently used data and instructions, and a dedicated running task implies the high chance of cache hit and is good for performance.

    If one core is shared to run multiple software threads, the task switch may imply that the existing data in cache is no longer valid for the new task to run. This means more cache miss for a new task, thus impacting the performance negatively due to the long memory latency to load data from DRAM. In NUMA architecture, this may become more complicated; in the worst case, the task switch involves the across NUMA data, as the task may be scheduled from socket 0 to socket 1, adding the further negative impacts to the data in L3 cache.

    Software developers need to take affinity and NUMA into consideration. Generally, if the software is hungry for performance, then leveraging the benefits of affinity configuration can help; otherwise, the Linux OS scheduler does a wonderful job, and the kernel has the advanced algorithm for the task scheduling. Here are a few specific suggestions.

    Computation-intensive workload: Massive calculation is generally found in scientific computing and theoretical calculation. A common indicator is the high CPU utilization ratio.

    Latency-sensitive and high determinism workload CPU affinity helps to address the real-time (latency-sensitive) need. This ensures the dedicated resource to avoid the scheduling disruption. Avoiding resource sharing is a good way to achieve high determinism.

    System performance tuning: Tuning the complex software is another reason why we are interested in CPU affinity. The ideal goal is to allow the system performance scale with more core counts.

3. CPU (Core) Isolation

Affinity binds the task to the assigned cores, but the Linux scheduler can load other tasks to the assigned cores. To further reduce the impact of adding other task's (owned by Linux OS) load to occupy the assigned cores, there is an option to inform Linux scheduler of which cores are available to be used, and which cores are not available for the shared use.

Linux provides a boot parameter: `isolcpus`. Assuming a server has been installed with four cores, it is possible to set up the boot parameters `isolcpus=2, 3` before the system launches. Then Linux system will assume the cores (representing by 2, 3) are no longer available, and note that the Do Not Use is not absolute. You can still use the taskset command to specify which programs can run on the core after the system boot. The steps are as follows:

Command: `vim /boot/grub2.cfg`

Add the isolcpus parameter `isolcpu=2,3` to the Linux kernel boot parameters

Command: `cat /proc/cmdline`

Check the boot parameters after the system restarts

```
BOOT_IMAGE=/boot/vmlinuz-3.17.8-200.fc20.x86_64
root=UUID=3ae47813-79ea-4805-a732-21bedcbdb0b5 ro LANG=en_
US.UTF-8 isolcpus=2,3
```

4. CPU (Core) Utilization, 100%?

Using affinity and core isolation, the assigned cores can be exclusively used by a specific software task. The dedicated cores assignment is the extreme and effective approach; this approach is selected as one of DPDK design principles; a core is often used with endless loop for packet processing, which also appears with 100% CPU utilization.

This is a controversial idea for many new users. In the old embedded world, 100% core utilization is often perceived as a software bug because the software is running in the endless loop. I worked on the embedded software development project, and there was a "watchdog" process, which requires to be fed periodically, and if it is not done in time, a hardware reset will trigger. It is designed as the system recovery mechanism to avoid the system running in deadlock.

In the earlier years, CPU only had one core, and 100% utilization was not acceptable as the other tasks also needed the execution resource. Now CPU comes with many cores, and using a small portion of cores with 100% utilization is a way to realize the performance critical goal, as there are still cores available for other tasks.

3.1.4 Core Pinning in DPDK

A thread in DPDK (EAL (environment abstraction layer) thread) is based on Linux pthread interface. It is governed by a Linux kernel scheduler, and it supports the pre-emptive threading model. DPDK suggests to create multiple threads to make good use of the multicore architecture. Software thread in DPDK is often assigned to the dedicated cores. It is designed to minimize the overhead of task switching. DPDK makes good use of the CPU pinning and CPU isolation. Linux scheduler does not manage the hardware cores that have been assigned to DPDK threads.

A DPDK thread can be used as either a DPDK main control thread or many data plane slave threads. The control thread, which is generally known as the master thread, is responsible for DPDK system configuration in order to set up the packet processing environment according to the command options and enable the data plane thread ("slave" cores). Most data plane threads are either distributed or running in parallel, and each thread occupies a specific logical core for processing packets.

DPDK samples are served as software reference. The system designer makes the final architectural decision on how to apply the thread model, CPU isolation, and also whether to implement a model to choose or avoid 100% CPU utilization. It is important to highlight that DPDK supports not only the poll mode driver, but also the interrupt mode driver. Poll mode is the most famous model for the extreme performance benefit. The interrupt mode is also gaining adoption in the cloud use case, where cloud network I/O requirements are not as stringent as telecom infrastructure.

1. EAL lcore

   In DPDK context, the `lcore` is equal to EAL pthread, which is realized on the pthread (Linux/FreeBSD) modules. EAL thread is created and managed by the task specified by the `remote _ launch` function. In each EAL pthread, it will be assigned with `_ lcore _ id` as the specific logical core, which comes from the initialization parameter "`-c`". EAL subsystem initializes the assigned lcore with the software entry function. This will create the actual mapping between lcore and EAL thread. Here is a close look into `rte _ eal _ init`, which includes the lcore initialization with software entry function registration in DPDK.

   Initialization

   1. Call `rte _ eal _ cpu _ init()` function; it will discover the available CPU cores in the current system, and build the socket and core relationship by reading the system information under `/sys/devices/ system/cpu/`.
   2. Call `eal _ parse _ args()` function, parse the "`-c`" to get the coremask, assign the usable cores, and set the first core as the "**master**" core.
   3. Create a thread for each "**slave**" core and call the `eal _ thread _ set _ affinity ()` to bind software entry function with hardware core. The execution thread is `eal _ thread _ loop()`. The main body of the `eal _ thread _ loop()` is a "while" loop, which calls different modules to register to the callback function under `lcore _ config[lcore _ id].f`.

   ```
   RTE _ LCORE _ FOREACH _ SLAVE(i) {
           /*
           * create communication pipes between master
           thread
   ```

```
* and children
*/
if (pipe(lcore _ config[i].pipe _ master2slave) <0)
     rte _ panic("Cannot create pipe\n");
if (pipe(lcore _ config[i].pipe _ slave2master) <0)
     rte _ panic("Cannot create pipe\n");
lcore _ config[i].state = WAIT;

/* create a thread for each lcore */
ret = pthread _ create(&lcore _ config[i].thread _
id, NULL,
    eal _ thread _ loop, NULL);
if (ret != 0)
     rte _ panic("Cannot create thread\n");
}
```

Registration

The master core needs to call rte _ eal _ mp _ remote _ launch(), and in the case of l2fwd example (its source code can be found under./examples/l2fwd/main.c), the entry function is specified as l2fwd _ launch _ one _ lcore().

rte_eal_mp_remote_launch(l2fwd_launch_one_lcore, NULL, CALL_ MASTER);

The slave cores will execute to a function point known as eal _ thread _ loop(), which is later jumped to the entry point l2fwd _ launch _ one _ lcore().Often this is the user-defined entry point. The slave core will run from here, as DPDK does not mandate the slave core's behavior, giving decision up to the software program. In many DPDK examples, the slave thread is running as an endless loop. The whole master and slave core initialization registration procedure can be summarized in Figure 3.3.

2. Lcore Affinity

By default, a software thread is assigned to the logical core (lcore) to avoid performance impact of thread migration. Using the affinity approach, the performance is enhanced at the cost of software flexibility; it is a design trade-off.

In the real network, the traffic isn't predictable as high or low, as it may follow the tide effect. In the daytime, the mobile and office-related Internet traffic are high, and in the late afternoon and evening, the residential Internet traffic will increase from low to high until midnight. At nighttime, the network traffic is significantly reduced.

It is unnecessary to apply the same number of cores for low or high traffic scenario. On-demand core assignment is a flexible and smart way of using resources. The application is responsible for resource management. The scale-up and scale-down model can help to deliver a networking infrastructure with the energy efficiency principle.

The below option is to assign a set of core for DPDK EAL pthread.

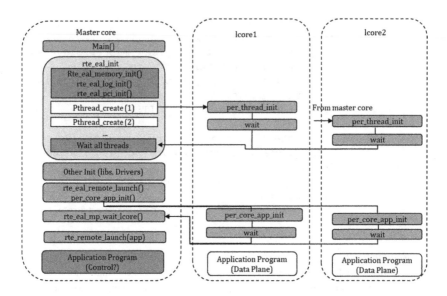

FIGURE 3.3   Lcore initialization and distribution of execution tasks.

The format is as follows:

```
--lcores='<lcore _ set>[@cpu _ set][,         <lcore _ set>[@
cpu _ set],...]',
```

where "lcore _ set" and "cpu _ set" can be a number, range, or group. The numerical value is "digit([0-9]+)", the scope is "<number>-<number>", and the group is "(<number|range>[, <number|range>,...])". If the value of "@cpu _ set" is not specified, it is by default the value of "lcore _ set". This option is compatible with that of "-l", one of the corelist options.

For example, "--lcores='1,2@(5-7),(3-5)@(0,2),(0,6),7-8'" which means start 9 EAL thread;

```
lcore 0 runs on cpuset 0x41 (cpu 0,6);
lcore 1 runs on cpuset 0x2 (cpu 1);
lcore 2 runs on cpuset 0xe0 (cpu 5,6,7);
lcore 3,4,5 runs on cpuset 0x5 (cpu 0,2);
lcore 6 runs on cpuset 0x41 (cpu 0,6);
lcore 7 runs on cpuset 0x80 (cpu 7);
lcore 8 runs on cpuset 0x100 (cpu 8).
```

This command option and related APIs (rte _ thread _ set/get _ affinity()) add a little more design flexibility. It provides dynamic resource allocation with a single core or a set of cores.

As said before, it is possible to allow multiple software threads to run on the same logical cores. It should be noted that the scheduling of multiple preemptive tasks on the same core need to be cautious, as there are design assumptions and restrictions when integrating with the non-preemptive library in DPDK. Take `rte _ ring` as an example:

1. In `single-producer/single-consumer` mode, the scheduling is not affected and can be used normally.
2. In `multiple-producer/multiple-consumer` mode with the pthread scheduling strategy being `SCHED _ OTHER`, it can be used but its performance may be affected.
3. If `multiple-producer/multiple-consumer` mode with the pthread scheduling strategy being `SCHED _ FIFO` or `SCHED _ RR`, it is not recommended for use; otherwise, a deadlock may occur.

3. Support for User-Created Threads

DPDK provides a method to create the pthread to run on assigned lcore, and it also allows Linux pthreads to spawn without using DPDK APIs. In fact, those Linux threads (user-created) can call some DPDK APIs. For the user-created pthread, its lcore_id is set as `LCORE _ ID _ ANY`. As default, it can run on any logical cores on Linux system, the same as any threads spawned on Linux. DPDK provides a very thin layer to establish an EAL for the optimized packet processing; the fundamental functions rely on Linux OS.

The user-created pthread can invoke most of the existing DPDK libraries without any negative effects, but a few DPDK libraries may not be recommended to use, such as `rte _ timer` and `rte _ mempool`. Take `rte _ mempool` as an example. In a user-created pthread, the mempool cache of per-core cannot be activated, which will have some impacts on the optimal performance. This will be explained in later chapters which focus on multithreading, more relevant information via online documentation.

4. Effective Core Utilization

DPDK performance is largely decided by the assigned logical core count. To deal with the increased or decreased network traffic, the system designer is responsible for the flexible way to scale up and down. DPDK process/threads are actually the common Linux process/pthread. Linux Cgroup (Control Group) can limit, record, and isolate the physical resources (CPU, memory, I/O, etc.) used by Linux process groups. Just like any other application, DPDK's resource consumption can also be managed by Linux Cgroup.

3.2 INSTRUCTION CONCURRENCY AND DATA PARALLELISM

We have discussed the performance scaling by using more cores. Now lets' shift focus at core level; there are additional optimization opportunities for software programmers.

3.2.1 INSTRUCTION CONCURRENCY

The modern multicore processor adopts the superscalar architecture to increase the instructions' concurrency. It can complete multiple instructions within a clock cycle. The Skylake server core microarchitecture pipeline is illustrated in Figure 3.4. Each

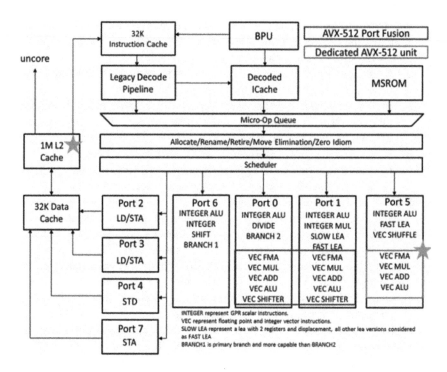

FIGURE 3.4 Skylake core architecture and pipeline.

core scheduler has eight ports, with four ALUs that can run in parallel. This provides a high-level picture to achieve high instruction parallelism. The data load/store ports are another indicator of the possibility of gaining data parallelism

Software developers cannot write the explicit code to control each individual unit within the core, Intel® Xeon CPU is designed with out-of-order execution, and Intel® Atom CPU is designed with in-order execution. Intel® recommends the software programming techniques, which drives the high parallelism execution, and is described by *x86 Architecture Optimization Manual* in great detail.

3.2.2 SIMD

Single-instruction multiple data (SIMD) improves performance by applying the instruction concurrency on the dimension of data width, and over the years, CPU has adapted to support the computation on the registers with 128-bit, 256-bit, and 512-bit width. Multiple data define a specific width as a data unit, and multiple data units are operated independently. "Single instruction" refers to the fact that for such a multiple data set, an instruction operation can be applied to all data units. SIMD is considered as a vectoring operation method. As shown in Figure 3.5, there are two data sets, as four data units for group (X1, X2, X3, X4 and Y1, Y2, Y3, Y4) are given, and the same operation is applied to the corresponding data unit pairs (X1 and Y1, X2 and Y2, X3 and Y3, X4 and Y4). The final result is concluded in the final four data units (Figure 3.5).

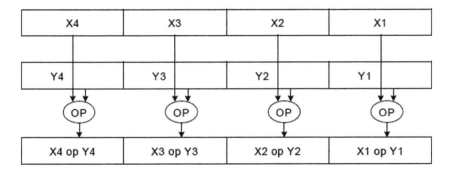

FIGURE 3.5 Example of SIMD operation.

Level	Capacity / Associativity	Line Size (bytes)	Fastest Latency[1]	Peak Bandwidth (bytes/cyc)	Sustained Bandwidth (bytes/cyc)	Update Policy
First Level Data	32 KB/ 8	64	4 cycle	96 (2x32B Load + 1*32B Store)	~81	Writeback
Instruction	32 KB/8	64	N/A	N/A	N/A	N/A
Second Level	256KB/4	64	12 cycle	64	~29	Writeback
Third Level (Shared L3)	Up to 2MB per core/Up to 16 ways	64	44	32	~18	Writeback

FIGURE 3.6 Skylake server microarchitecture—cache details

SIMD uses the register sets with much wider than the general-purpose registers (GPRs) in order to hold the wide data units. It has evolved with 128-bit XMM registers, 256-bit YMM registers, and 512-bit ZMM registers, but the GPRs are still 64-bit width; this is also why the current CPU is still named as 64-bit (ia64). A direct benefit of using SIMD instruction is maximizing the data parallelism; each cache line is about 64 bytes, known as 512-bit width (Figure 3.6).

DPDK is designed for I/O-intensive scenarios. SIMD has been proven to improve the performance significantly, one example being the batched packet processing, such as in PMD's Rx/Tx functions where adjacent descriptors can be processed together by AVX (advanced vector extensions) instructions. For example, each descriptor size of the PMD is 16 bytes, and then four descriptors can be processed together (64 bytes) with one AVX512 instruction.

Another example is the memory copy in DPDK. A memory copy is a very simple operation without a sophisticated algorithm. For the aligned data moving from one place to another, we can use the widest load and store instructions supported by the platform (512 bits for Skylake). To complete the copy of 4K cached data, it may take 261 cycles using AVX (128 bits), take 132 cycles using AVX2 (256 bits), and take 80 cycles using AVX512 (512 bits).

Additionally, the nonaligned memory access operation generally requires more clock cycles, and the source code will be described in more detail. If a user has a design choice, use the aligned memory address, and consider the SIMD use in mind.

```
Performance counter stats for './stream_c.exe':

       229.872935 task-clock                #   0.996 CPUs utilized
      626,676,991 cycles                     #   2.726 GHz
      525,543,766 instructions               #   0.84  insns per cycle
       18,587,219 cache-references           #  80.859 M/sec
        6,605,955 cache-misses              #  35.540 % of all cache refs

      0.230761764 seconds time elapsed
```

FIGURE 3.7 Linux/IPC example. (https://developers.redhat.com/blog/2014/03/10/deter
mining-whether-an-application-has-poor-cache-performance-2/.)

```
/**
 * Copy 64 bytes from one location to another,
 * locations should not overlap.
 */
static __rte_always_inline void
rte_mov64(uint8_t *dst, const uint8_t *src)
{
        __m512i zmm0;

        zmm0 = _mm512_loadu_si512((const void *)src);
        _mm512_storeu_si512((void *)dst, zmm0);
}
```

3.3 CONCLUSION

Inside each CPU core, hardware-based execution parallelism such as instruction-level parallelization (ILP) and SIMD helps the software performance. IPC is an important indicator to measure the software efficiency. Linux "Perf" is a software utility that can profile and measure how applications perform. The above example indicates "IPC = 0.84". In the sample of Skylake processor, four execution units are available, thus limiting the upper bound of "IPC = 4". The higher IPC, the more efficiently software is designed (Figure 3.7).

A well-thought design needs to split the software task into many parallel tasks, so each task can run separately on the multicore systems; this design principle can provide the future processor opportunity to gain more performance, as more cores will be added in the new processor generation. However, the software parallelism is bound if there is a need for data synchronization among multiple cores. The data synchronization is expensive, and the lock-free structure will be introduced in Chapter 4.

FURTHER READING

1. Columbia University Material, History of Processor Performance http://www.cs.columbia.edu/~sedwards/classes/2012/3827-spring/advanced-arch-2011.pdf.

4 Synchronization

Frank Liu
NetEase

Heqing Zhu
Intel®

CONTENTS

Multicore architecture creates a high parallelism hardware platform environment; data sharing is a critical need among cores; packet processing requires the packets can be shared from one core to another for a high performance system; Essentially this is similar to the classic consumer and producer model. In the load balancing pipeline scenario, the packet sharing can go from a single producer to multiple consumers model, or even multiple producers to multiple consumers model. Software designers need to use the effective data-sharing methods among CPU cores such as data synchronization and critical resource protection. This chapter describes the rte_ring, rte_atomic, lock-free, memory barrier concept.

4.1   ATOMIC AND MEMORY BARRIER

An atom refers to "the smallest particle that cannot be further divided". From the computing perspective, atomic operation refers to "one operation cannot be interrupted". In a multiple threads' software program, one thread completes the operation, whereas the other threads will see that the operation is completed instantaneously. Atomic operation is the foundation for the synchronization, and atomic memory access is important, as all cores share the memory.

4.1.1   MEMORY BARRIER

Memory barrier, also known as memory fence, enforces the memory access (load/store) in order. Memory load or store can be expensive as the access latency can go up to hundreds of cycles in the context of DRAM access. Intel® Xeon-based CPU has designed with out-of-order execution; software compiler can help with the program optimization attempt. CPU and compiler may find ways to execute program with less memory access, which may lead to program reorder, including the reorder of memory access. The impact is the program execution order is different than the software code order. If the shared data is already loaded from memory to cache, the memory read/modify/write (RMW) operation can happen in the cache, even used by other cores; this means less latency due to the caching benefits. In some cases, it is okay to allow the program reorder, but not always.

Manage the shared data's memory use to follow the exact program order; x86 supports LFENCE, SFENCE, and MFENCE instructions to realize the memory barriers.

- *LFENCE* is a memory load fence; this ensures a serialization operation on all the previous memory loads that are globally visible.
- *SFENCE* is a memory store fence; it only enforces that all previous memory stores are globally visible.
- *MFENCE* is the combination to enforce that all the previous memory load and store operations are globally visible.

In DPDK, the packet is received in the memory buffer; it needs to be shared from NIC (device) to core, from core to core, and from host to guest (such as virtualization or container tenants). It does use memory barrier in a few places like NIC PMD.

4.1.2   ATOMIC OPERATION

In the old days of a single-processor (=core) system, a single instruction can be regarded as "atomic operation", because interrupt can only occur among instructions, if there are no other tasks running in parallel. The memory operation like "INC" is indeed memory RMW; it is "atomic" in a single-task environment. In a multicore system, if the instruction is not performed on the shared data, such an instruction is atomic, because there is no other access to the data. When multiple cores are trying to access the shared data, even for a simple "INC", it is still translated into actions like the memory RMW, but for all involved cores. When multiple cores are accessing the same data in memory, the data might be cached by a core; the cached data is also used by other cores; the remaining cores will be updated automatically if there is any data change; x86 is designed with cache coherence to guarantee this, but it comes at a cost of using more cycles to achieve the data coherency.

The core can perform reads and writes to the aligned bytes, words, double words, and four bytes atomically; for unaligned bytes, words, double words, and four bytes, their reads and writes might be atomic if the memory load/store happened in the same cache line, because CPU always loads/stores data in the size of a cache line. But if the data resides in the multiple cache lines, this becomes complicated as a result of multiple load/store operations. A simple solution is to ensure that the shared data stays in the same cache line, which avoid the unnecessary data spreading over multiple cache lines. A program directive like " _ _ rte _ cache _ aligned" can ask the compiler to enforce the cache line alignment on the specified data structure. This works for the small data structure, whose size is less than 64 bytes (x86).

x86 supports "LOCK" instruction prefix; it was a bus lock; other cores will not access the memory if "LOCK" signal is asserted. This helps the relevant instructions to achieve the atomic memory access. This is not free as other cores are now waiting for this to complete first. "Compare and swap (CAS)" is a common way for multi-threading to achieve synchronization. x86 implements this as an instruction— "compare and exchange (CMPXCHG)". LOCK prefix is used together on a multicore system. This is the foundation to implement the lock or is used for lockless data structure design. Together with lock, "CMPXCHG" instruction is used for lockless data structures; the lockless queue is performant way for the multicore to exchange the data.

4.1.3   LINUX ATOMIC OPERATION

Software atomic operation relies on hardware atomic operation. Linux kernel supports the atomic operations for simple flags, atomic counters, exclusive ownership, and shared ownership in the multi-threaded environment. It provides with two sets of software interfaces: One is the integer level, and the other is the bit level.

4.1.3.1   Atomic Integer Operation

A common use of the atomic integer operation is to implement the statistics counter. The atomic operation for integers can go with the data type "atomic _ t". The atomic function is suggested to use the "atomic _ t" on input/output parameters. The compiler will not try to optimize here. Generally, atomic operation is an inline function and implemented by the inline assembly instructions. If a function itself is atomic, it will be often defined as one macro. Atomicity can ensure the program won't be interrupted when the instruction is executed completely or not executed at all. The sequence ensures that even if two or more instructions are in the independent execution thread or independent processing units, the original order will be kept in execution time.

4.1.3.2   Atomic Bit Operation

In many use cases, the atomic bit operation is the access to a 1-word-long memory region, 0–31 on a 32-bit machine, and bits 0–63 on a 64-bit machine. In Linux kernel, the atomic bit operation is defined as follows:

```
"include\linux\types.h"
"arch\x86\include\asm\bitops.h".
```

Here are Linux API samples on atomic operation. For the integer atomic operation functions, as shown in Table 4.1, the following addition operations have the corresponding subtraction operations (Table 4.2).

TABLE 4.1

Integer Atomic Operation

Atomic Integer Operation	Functions
ATOMIC_INIT(int i)	When one atomic_t variable is claimed, it shall be initialized as i
int atomic_read(atomic_t *v)	Atomically read the integer variable v
void atomic_set(atomic_t *v, int i)	Atomically set the value of v as i
void atomic_add(int i, atomic_t *v)	Atomically add i tov
void atomic_sub(int i, atomic_t *v)	Atomically subtract i from v
void atomic_inc(atomic_t *v)	Atomically add i to v
void atomic_dec(atomic_t *v)	Atomically subtract 1 from v
int atomic_sub_and_test(int i, atomic_t *v)	Atomically subtract 1 from v. If the result is 0, returns true; otherwise, returns false.
int atomic_add_negative(int i, atomic_t *v)	Atomically add i to v. If the result is negative, returns true; otherwise, returns false.
int atomic_dec_and_test(atomic_t *v)	Atomically subtract 1 from v. If the result is 0, returns true; otherwise, returns false.
int atomic_inc_and_test(atomic_t *v)	Atomically add 1 to v. If the result is 0, returns true; otherwise, returns false.

TABLE 4.2

Bit Atomic Operation

Atomic Bit Operation	Functions
void set_bit(int nr, void *addr)	Atomically set No. nr of the object where addr points
void clear_bit(int nr, void *addr)	Atomically clear No. nr of the object where addr points
void change_bit(int nr, void *addr)	Atomically flip No. nr of the object where addr points
int test_and_set_bit(int nr, void *addr)	Atomically set No. nr of the object where addr points and return the original value
int test_and_clear_bit(int nr, void *addr)	Atomically clear No. nr of the object where addr points and return the original value
int test_and_change_bit(int nr, void *addr)	Atomically flip No. nr of the object where addr points and return the original value
int test_bit(int nr, void *addr)	Atomically return No. nr of the object where addr points

4.1.4   DPDK ATOMIC OPERATION

Atomic operation is about data access, and data sits in the memory. The atomic interface is mostly about the atomic memory access. DPDK implements its own memory barriers and atomic interfaces. Source files are located at

- lib/librte _ eal/common/include/generic/rte _ atomic.h
- lib/librte _ eal/common/include/arch/x86/rte _ atomic.h (x86 specific)

4.1.4.1   Memory Barrier API

It keeps the memory load/store instructions issued before the memory barriers are global visible (for all cores).

- *rte_mb()*: Memory barrier read-write API, using MFENCE.
- *rte_wmb()*: Memory barrier write API, using SFENCE.
- *rte_rmb()*: Memory barrier read API, using LFENCE.

In the below example of DPDK, vhost is running in the "host" system; it needs to notify "guest" on the interrupt arrival; the memory barrier API is added to send the notification. Generally, host and guests are running in the different cores.

```
static int eth_rxq_intr_enable(struct rte_eth_dev *dev,
uint16_t qid)
{
    struct rte_vhost_vring vring;
    struct vhost_queue *vq;
    int ret = 0;
    vq = dev->data->rx_queues[qid];
    if (!vq) {
            VHOST_LOG(ERR, "rxq%d is not setup yet\n", qid);
            return -1;
    }

    ret = rte_vhost_get_vhost_vring(vq->vid, (qid
    if (ret < 0) {
            VHOST_LOG(ERR, "Failed to get rxq%d's vring\n",
            qid);
            return ret;
    }

    VHOST_LOG(INFO, "Enable interrupt for rxq%d\n", qid);
    rte_vhost_enable_guest_notification(vq->vid, (qid << 1) +
    1, 1);
    rte_wmb();
    return ret;
}
```

4.1.4.2 Atomic API

The atomic data types and interfaces are supported as 16, 32, and 64 bits long.

```
typedef struct {
        volatile int16_t cnt; /**< An internal counter value. */
} rte_atomic16_t;

typedef struct {
        volatile int32_t cnt; /**< An internal counter value.
*/
} rte_atomic32_t;

typedef struct {
        volatile int64_t cnt; /**< An internal counter value. */
} rte_atomic64_t;
```

The atomic API is implemented with "LOCK" instruction prefix. The below source code of rte_atomic64_add() is an example.

```
static inline void
rte_atomic64_add(rte_atomic64_t *v, int64_t inc)
{
        asm volatile(
                    MPLOCKED
                    "addq %[inc], %[cnt]"
                    : [cnt] "=m" (v->cnt)      /* output */
                    : [inc] "ir" (inc),        /* input */
                    "m" (v->cnt)
                    );
}
```

The atomic INC/ADD interface is often used for packet counter statistics; **virtio_
xmit** is to send packets to guest; the counter may be accessed in the multicore system. When statistics feature is enabled, the packet counter is incremented through atomic operation.

```
static __rte_always_inline void
virtio_xmit(struct vhost_dev *dst_vdev, struct vhost_dev
*src_vdev,
                struct rte_mbuf *m)
{
    uint16_t ret;
    if (builtin_net_driver) {
        ret = vs_enqueue_pkts(dst_vdev, VIRTIO_RXQ, &m, 1);
    } else {
        ret = rte_vhost_enqueue_burst(dst_vdev->vid, VIRTIO_
        RXQ, &m, 1);
    }
```

```
    if (enable_stats) {
        rte_atomic64_inc(&dst_vdev->stats.rx_total_atomic);
        rte_atomic64_add(&dst_vdev->stats.rx_atomic, ret);
        src_vdev->stats.tx_total++;
        src_vdev->stats.tx += ret;
    }
}
```

Atomic "compare and set" operation is heavily used for lock and lockless design. As said before, DPDK implemented this with LOCK prefix and CMPXCHG instructions together.

```
static inline int
rte_atomic32_cmpset(volatile uint32_t *dst, uint32_t exp,
uint32_t src)
{
        uint8_t res;
        asm volatile(
                MPLOCKED
                "cmpxchgl %[src], %[dst];"
                "sete %[res];"
                : [res] "=a" (res),           /* output */
                [dst] "=m" (*dst)
                : [src] "r" (src),            /* input */
                "a" (exp),
                "m" (*dst)
                : "memory");                  /* no-clobber
list */
        return res;
}
```

A complete list of atomic interfaces can be found in the source code.

4.2 RWLOCK

The RW (Reader Writer) lock—rwlock—is a lock for resource protection. If it is read-only access, the shared memory will not be modified, and the multiple core (reader) access is safe for parallel access. But write access is different as data will be modified; the access needs to be exclusive. Compared to the spinlock, the rwlock improves the read concurrency in the multicore system. The write operation is exclusive. This means that only a write operation is allowed, while others need to delay the access until the write is completed.

The RW spinlock is characterized as follows:

- Resources between read locks are shared. When a thread has a read lock, other threads can share the access by reading only.
- Write locks are mutually exclusive. When a thread has a write lock, other threads can't share it either by reading or by writing.

- Read and write locks are mutually exclusive. When a thread has a read lock, other threads can't access it by writing.

The shared resource can have many formats: a simple variable or a complicated data structure. The shared resource is associated with a unique rwlock, and the access to the shared resource follows the below steps:

1. Request the lock.
2. Access the shared resource after the lock is acquired.
3. Release the lock.

RW lock must satisfy the following three requirements:

1. *Mutex*: A reader and a writer cannot access the shared resource (i.e., acquire the lock) simultaneously at any time.
2. *Reader concurrency*: Multiple readers may access the shared resource simultaneously.
3. *No deadlock*: If a thread is waiting for the lock, the lock shall be owned by someone.

The rwlock should be used with caution; it is suggested to be used for a short period; if the access to the shared resource is completed, then the lock needs to be released as soon as possible. A waiting thread shall not enter sleep because the thread entering sleep and then wakeup can take many cycles, considering that this is expensive, which will increase the wait cost for the lock acquisition. Therefore, the busy-waiting mode is a good option. A thread requesting a lock will repeatedly check whether any other thread has released the lock; it shall not enter the sleep mode.

4.2.1 Linux API

Linux API has the RW lock interface in both kernel space and user space. Tables 4.3 and 4.4 lists the specific software interface. RW lock implementation can be architecture-specific; please refer to Linux source code for details.

TABLE 4.3
Linux RW Locks in Kernel Space

Interface	Description
rwlock_init()	Initializes a rwlock_t instance
read_lock()	Acquires the specified reader lock
read_lock_irq()	Disables the local interrupt, and acquires the specified reader lock
read_lock_irqsave()	Stores the current state of the local interrupt, disables the local interrupt, and acquires the specified reader lock
read_unlock()	Releases the specified reader lock

(Continued)

TABLE 4.3 (*Continued*)
Linux RW Locks in Kernel Space

Interface	Description
read_unlock_irq()	Releases the specified reader lock, and activates the local interrupt
read_unlock_irqrestore()	Releases the specified reader lock, and returns the local interrupt to the state before specifying
write_lock()	Acquires the specified writer lock
write_lock_irq()	Disables the local interrupt, and acquires the specified writer lock
write_lock_irqsave()	Stores the current state of the local interrupt, disables the local interrupt, and acquires the specified writer lock
write_unlock()	Releases the specified writer lock
write_unlock_irq()	Releases the specified writer lock, and activates the local interrupt
write_unlock_irqrestore()	Releases the specified writer lock, and returns the local interrupt to the state before specifying
write_trylock()	Attempts to acquire the specified writer lock; if the writer lock is not available, returns to a value other than 0

TABLE 4.4
Linux RW Locks in User space (POSIX)

Interface	Description
pthread_rwlock_init()	Initializes a rwlock object
pthread_rwlock_destroy()	Destroys the rwlock object
pthread_rwlock_rdlock()	Requests to acquire the rwlock object for reading; the caller will block until acquired
pthread_rwlock_tryrdlock()	Requests to acquire the rwlock object for reading; returns the success or failure immediately without block
pthread_rwlock_timedrdlock()	Requests to acquire the rwlock object for reading; the caller will block until timeout
pthread_rwlock_unlock()	Unlocks the rwlock object
pthread_rwlock_wrlock()	Locks the rwlock for writing; the caller will block until acquired
pthread_rwlock_trywrlock()	Locks the rwlock for writing; the caller will not block
pthread_rwlock_timedwrlock()	Locks the rwlock for writing; the caller will block until timeout

4.2.2 DPDK API

DPDK is mainly about a user space library; it does not use the standard POSIX rwlock interface; it has its own simple implementation. The lock object itself is 32-bit integer. The lock software interface is based on the atomic operations using "compare and exchange"; the caller block mechanism is implemented with "PAUSE" instruction within the endless loop mechanism. Read concurrently is based on the rwlock's internal counter increment and decrement.

```
typedef struct {
volatile int32_t cnt;
/**< -1 when W lock held, > 0 when R locks held. */
} rte_rwlock_t;

static inline void
rte_rwlock_read_lock(rte_rwlock_t *rwl)
{
    int32_t x;
    int success = 0;

    while (success == 0) {
        x = __atomic_load_n(&rwl->cnt, __ATOMIC_RELAXED);
        /* write lock is held */
        if (x < 0) {
            rte_pause();
            continue;
        }
        success = __atomic_compare_exchange_n(&rwl->cnt, &x, x
+ 1, 1,
                        __ATOMIC_ACQUIRE, __ATOMIC_RELAXED);
    }
}
```

Table 4.5 lists the DPDK-based rwlock interface. It does not support the timeout mechanism (Table 4.5).

TABLE 4.5

DPDK rwlock API

Interface	Description
rte_rwlock_init()	Initializes the rwlock object
rte_rwlock_destroy()	Destroys the rwlock object
rte_rwlock_read_lock()	Locks the rwlock object for reading; the caller will block until acquired
rte_rwlock_trylock()	Locks the rwlock object for reading; returns the success or failure immediately without block
rte_rwlock_read_unlock()	Unlocks the rwlock object for reading; decrements the lock counter. Supports the concurrent readers
rte_rwlock_write_lock()	Locks the rwlock for writing; the caller will block until acquired
rte_rwlock_write_trylock()	Locks the rwlock for writing; the caller will not block
rte_rwlock_write_unlock()	Unlocks the rwlock object; zero the counter

The DPDK-based "rwlock" has the following use cases in DPDK for protecting the shared resource:

- Memory zone and segments used the rwlock when looking for the idle memory segments—LPM (longest prefix match) routing table management;
- Table management such as Hash and IPsec Security Association;
- IOTLB management by vhost;
- Some PMD configuration interface.

Here is a code example for memzone lookup; the read lock is applied; for the memory zone initialization, the write lock is used.

```
/* Apply the read lock as concurrent reading */
const struct rte_memzone *
rte_memzone_lookup(const char *name)
{
        struct rte_mem_config *mcfg;
        const struct rte_memzone *memzone = NULL;
        mcfg = rte_eal_get_configuration()->mem_config;
        rte_rwlock_read_lock(&mcfg->mlock);
        memzone = memzone_lookup_thread_unsafe(name);
        rte_rwlock_read_unlock(&mcfg->mlock);
        return memzone;
}

/* Apply the write lock for initialization (write) */
int rte_eal_memzone_init(void)
{
        struct rte_mem_config *mcfg;
        int ret = 0;

        /* get pointer to global configuration */
        mcfg = rte_eal_get_configuration()->mem_config;

        rte_rwlock_write_lock(&mcfg->mlock);

        if (rte_eal_process_type() == RTE_PROC_PRIMARY &&
          rte_fbarray_init(&mcfg->memzones, "memzone",
          RTE_MAX_MEMZONE, sizeof(struct rte_memzone))) {
            RTE_LOG(ERR, EAL, "Cannot allocate memzone
list\n");
                ret = -1;
        } else if (rte_eal_process_type() == RTE_PROC_SECONDARY
&&
                rte_fbarray_attach(&mcfg->memzones)) {
                RTE_LOG(ERR, EAL, "Cannot attach to memzone
list\n");
                ret = -1;
        }
```

```
        rte_rwlock_write_unlock(&mcfg->mlock);
        return ret;
}
```

4.3   SPINLOCK

Spinlock is a locking mechanism to access a shared resource. Before acquiring the lock, the caller can stay in the busy-waiting mode. Once the lock is acquired, the access to shared resource will get started; when the access is completed, the spinlock can be released. In the previous section, write lock is a spinlock, which is used for the mutual exclusion scenario, but read lock is different, it is counter-based for concurrent access.

Spinlock is designed as a solution for the mutually exclusive use of shared resources; it is similar to the mutex lock, but there is a difference. For a mutex lock or a spinlock, only one owner can acquire the lock. However, if the lock is acquired by others,

- For the mutex lock, the caller will enter the sleep mode.
- For the spinlock, the caller will keep spinning—busy waiting.

Because the spinlock is busy waiting, it has no extra cost of context switching, so it is faster. It is worth noting that core is not doing anything useful in the busy wait period. The spinlock needs to be used with caution; it is easy to cause the deadlock if there is a repeated lock request from the same caller thread. Note that DPDK has not implemented its own mutex.

4.3.1   LINUX API

In the Linux kernel, spinlock implementation is specific to the CPU architecture. The inline assembly code is located in the relevant folder; more details can be found at the header file <asm/spinlock.h>. The typical use case is simple as follows:

```
spinlock_t lock = SPIN_LOCK_UNLOCKED;spin_lock(&lock);

Executing the critical section to access shared resources.
/* Critical section */

spin_unlock(&lock);
```

A subset of spinlock interfaces are summarized in Table 4.6.

When using the spinlock, the recursive request may lead to the deadlock. In the context of interrupt handling, the spinlock needs to be used more carefully. Generally, it is recommended to disable interrupt first, and this will avoid thread and interrupt competition. The API is available to disable interrupt before using the spinlock (Table 4.6).

TABLE 4.6
Linux Spinlock API

Interface	Description
spin_lock_init()	Initializes a spinlock object
spin_lock()	Acquires the spinlock
spin_lock_irq()	Disables interrupt and acquires the lock
spin_lock_irqsave()	Saves the current state of the interrupt, disables interrupt, and acquires the specified lock
spin_unlock()	Releases the specified lock
spin_unlock_irq()	Releases the specified lock and enables interrupt
spin_trylock()	Attempts to acquire the specified lock, return without busy waiting if it is acquired by others
spin_is_locked()	Queries if the lock is being acquired

TABLE 4.7
Linux spinlocks in user space (POSIX)

Interface	Description
pthread_spin_init()	Initializes a spinlock object
pthread_spin_destroy()	Destroys the spinlock object
pthread_spin_lock()	Requests to acquire the lock object; the caller will block until acquired
pthread_spin_tryrdlock()	Requests to acquire the lock object; returns the success or failure immediately without block
pthread_spin_unlock()	Releases the lock object

For Linux user space, Table 4.7 lists the POSIX spinlock interface.

4.3.2 DPDK API

Similar to rwlock, DPDK implemented its own spinlock; the source code is in the rte _ spinlock.h file; the spinlock interface is widely used with DPDK modules such as alarm, log, interrupt, memory sharing, and link bonding.

```
typedef struct {
volatile int locked; /**< lock status 0 = unlocked,
1 = locked */
} rte_spinlock_t;

static inline void
rte_spinlock_lock(rte_spinlock_t *sl)
{
        int exp = 0;

        while (!__atomic_compare_exchange_n(&sl->locked, &exp,
        1, 0,
```

```
              __ATOMIC_ACQUIRE, __ATOMIC_RELAXED)) {
              while (__atomic_load_n(&sl->locked,
              __ATOMIC_RELAXED))
                 rte_pause();
              exp = 0;
          }
}

/* Use example */
set_channel_status_all(const char *vm_name, enum channel_
status status)
{
...
rte_spinlock_lock(&(vm_info->config_spinlock));
mask = vm_info->channel_mask;
ITERATIVE_BITMASK_CHECK_64(mask, i) {
vm_info->channels[i]->status = status;
num_channels_changed++;
}
rte_spinlock_unlock(&(vm_info->config_spinlock));
return num_channels_changed;
}
```

Furthermore, DPDK implements the "recursive" spinlock. If a caller repeatedly requests the spinlock, it may run to the deadlock. The recursive lock can avoid this deadlock. The software will check if this is the same caller. If the caller has the lock already, it will increase the counter and the program can continue running (Table 4.8).

TABLE 4.8
DPDK spinlock API

Interface	Description
rte_spinlock_init()	Initializes a spinlock object, locked = 0
rte_spinlock_destroy()	Destroys the spinlock object
rte_spinlock_lock()	Requests to acquire the lock object; the caller will block until acquired. Busy waiting is based on "Pause" within the endless loop
rte_spinlock_trylock()	Requests to acquire the lock object; returns the success or failure immediately
rte_spinlock_is_locked()	Queries if spinlock in use
rte_spinlock_recursive_init()	Initializes a recursive spinlock object, locked = 0
rte_spinlock_recursive_destroy()	Destroys the recursive spinlock object
rte_spinlock_recursive_lock()	Requests to acquire the lock object; the caller will block until acquired. If it is the same caller, the counter will increment.
rte_spinlock_recursive_trylock()	Requests to acquire the lock object; returns the success or failure immediately. If it is the same caller, the counter will increment
rte_spinlock_recursive_unlock()	Releases the recursive spinlock; decrements the counter if it is the same caller.

```
/**
 * The rte_spinlock_recursive_t type.
 */
typedef struct {
        rte_spinlock_t sl; /**< the actual spinlock */
        volatile int user; /**< core id using lock, -1 for
unused */
        volatile int count; /**< count of time this lock has
been called */
} rte_spinlock_recursive_t;

static inline void
rte_spinlock_recursive_init(rte_spinlock_recursive_t *slr)
{
        rte_spinlock_init(&slr->sl);
        slr->user = -1;
        slr->count = 0;
}

static inline void
rte_spinlock_recursive_lock(rte_spinlock_recursive_t *slr)
{
        int id = rte_gettid();

        if (slr->user != id) {
                rte_spinlock_lock(&slr->sl);
                slr->user = id;
        }
        slr->count++;
}
```

4.4   LOCK-FREE

When lock is in use, the waiting cores are not utilized in an optimized way. Lock helps the data synchronization, but it forces the other cores to wait until the shared resource is available. The server platform has lots of core for software use. If software focuses on high performance, core-based scaling is important to achieve the high concurrency, the resource is ideally localized, and the data sharing needs to be as less as possible. But data sharing is often required, in this case, a lock-free mechanism can improve software performance. This enables more cores to work meaningfully in parallel. Performance scaling is a key design goal for DPDK, a multi-threading software that focuses on moving packets from cores to cores safely and fast.

To realize this, DPDK introduced its ring library; this is a specific optimization for packet processing workload, known as rte _ ring; it is based on FreeBSD's bufring mechanism. In fact, rte_ring supports the flexible enqueue/dequeue models for different cases.

- Single-producer (enqueue) and single-consumer (dequeue) model;
- Single-producer (enqueue) and multi-consumer (dequeue) model;

- Multi-producer (enqueue) and single-consumer (dequeue) model;
- Multi-producer (enqueue) and multi-consumer (dequeue) model.

The rte_ring library is often used (but not limited) together with rte_mbuf. As learned earlier, rte_mbuf is used for the packet buffer. So packet sharing can be done via rte_ring among cores, and the buffer pointers are objects in rte_ ring. The DPDK threads (cores) can act as producers or consumers, the actual packets stay in the same memory buffer, no actual memory copy is needed for each packet, but memory reference pointers are shared via rte_ring. Here, the enqueue/dequeue functions are indeed reading/writing rte_mbuf into rte_ring.

4.4.1   RTE _ RING

It is a simple data structure using DPDK-based memory zone concept, which has the optimized memory use and cache alignment; its source code is located in the folder lib/librte_ring/. The pad area is inserted to ensure cache line alignment as the producers and consumers are often running on the different cores (threads).

```
/* structure to hold a pair of head/tail values and other
metadata */
struct rte_ring_headtail {
        volatile uint32_t head; /**< Prod/consumer head. */
        volatile uint32_t tail; /**< Prod/consumer tail. */
        uint32_t single;        /**< True if single prod/cons */
};

struct rte_ring {
        char name[RTE_MEMZONE_NAMESIZE] __rte_cache_aligned;
        int flags;    /**< Flags supplied at creation. */
        const struct rte_memzone *memzone;

        uint32_t size;           /**< Size of ring. */
        uint32_t mask;           /**< Mask (size-1) of ring. */
        uint32_t capacity;       /**< Usable size of ring */

        char pad0 __rte_cache_aligned; /**< empty cache line */

        /** Ring producer status. */
        struct rte_ring_headtail prod __rte_cache_aligned;
        char pad1 __rte_cache_aligned; /**< empty cache line */

        /** Ring consumer status. */
        struct rte_ring_headtail cons __rte_cache_aligned;
        char pad2 __rte_cache_aligned; /**< empty cache line */
};
```

It consists of **prod** (producer) and **cons** (consumer); both have their own head/tail pointers. The ring size is decided at the creation phase, a memory reservation will happen from the DPDK-reserved memory zone, and the memory can be specified on the specific socket.

```
/* create the ring */
struct rte_ring *
rte_ring_create(const char *name, unsigned count, int
socket_id,
             unsigned flags)
```

count: Must be a power of 2.
socket_id: Optimized NUMA awareness. Use rte_socket_id() to select the caller's socket memory; or use SOCKET_ID_ANY for single-socket system.
flags: Initial flags for ring attributes such as single/multi-producer, single/multi-consumers.

4.4.2 ENQUEUE/DEQUEUE

Enqueue/dequeue APIs are available on a single object or multiple objects. The ring has been created for single- or multiple-producer/consumer flags. The software implementation can retrieve the ring flags and then enqueue/dequeue appropriately.

```
static __rte_always_inline int
rte_ring_enqueue(struct rte_ring *r, void *obj)
{
        return rte_ring_enqueue_bulk(r, &obj, 1, NULL) ? 0 :
        -ENOBUFS;
}

static __rte_always_inline int
rte_ring_dequeue(struct rte_ring *r, void **obj_p)
{
        return rte_ring_dequeue_bulk(r, obj_p, 1, NULL) ? 0 :
-ENOENT;
}

static __rte_always_inline unsigned int
rte_ring_enqueue_bulk(struct rte_ring *r, void * const *obj_
table, unsigned int n, unsigned int *free_space)

static __rte_always_inline unsigned int
rte_ring_dequeue_bulk(struct rte_ring *r, void **obj_table,
unsigned int n, unsigned int *available)
```

4.4.3 PRODUCER/CONSUMER

rte_ring has been created with single/multiple and producer/consumer flags; the internal enqueue/dequeue implementations are different; extra cautious is required

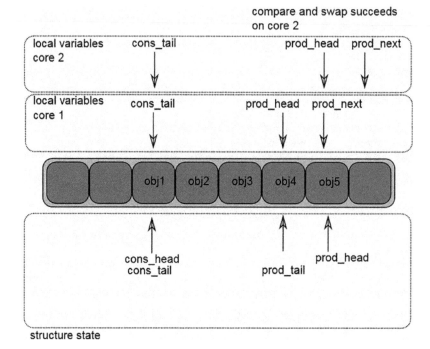

FIGURE 4.1   Ring internal for multiple producers and single consumer

to update the head/tail pointers; rings are shared resources in the multiple mode, in that case multiple cores (threads) may enqueue objects to the same ring simultaneously (Figure 4.1).

For the ring head update, the multiple producers or consumers can take advantage of "compare and set" function.

```
/* Ring Enqueue , Producer Head  */
{
   if (is_sp)  /* Single producer */
        r->prod.head = *new_head, success = 1;
   else   /* Multiple producers */
        success = rte _ atomic32 _ cmpset(&r->prod.head,*old_
head, *new_head);
}

/* Ring Dequeue, Consumer Head  */
{
   if (is_sc)  /* Single consumer */
        r->cons.head = *new_head,
        rte_smp_rmb();
        success = 1;
   else         /* Multiple consumers */
        success = rte _ atomic32 _ cmpset(&r->pcons.head,*old_
head, *new_head);
```

```
}

/* Ring Tail update, memory barrier and rte_pause */

static __rte_always_inline void
update _ tail(struct rte_ring_headtail *ht, uint32_t old_val,
uint32_t new_val, uint32_t single, uint32_t enqueue)
{
        if (enqueue)
                rte_smp_wmb();
        else
                rte_smp_rmb();
        /*
        * If there are other enqueues/dequeues in progress that
preceded us,
        * we need to wait for them to complete
        */
        if (!single)
                while (unlikely(ht->tail != old_val))
                        rte_pause();

        ht->tail = new_val;
}
```

4.5 CONCLUSION

Atomic, memory barriers are the fundamental concept for multi-threaded software; atomic interface is available on bits and integer level. Memory barrier ensures the correct order, whereas multiple cores need access to the shared memory. This leads to DPDK-based rwlock, spinlock implementation. Software program decides when to use memory barriers and atomic functions. rte_ring is known with lock-free design, it is the fastest way to move packets among cores, the packet movement can happen among cores without the actual packet memory copy and without using lock.

5 Forwarding

Yipeng Wang and Jasvinder Singh
Intel®

Zhe Tao
Huawei

Liang Ma and Heqing Zhu
Intel®

CONTENTS

Running the network functions purely in software is disruptive; it increases flexibility and agility of the infrastructure with the aim of reducing the total cost of ownership (TCO). The idea has been proved by the leading cloud service providers who are transforming their internal networking deployment. However, the transformation journey is not smooth. Indeed, there are challenges in a rapid rollout of the software network functions (often using open source) in the production systems. The functional requirements are easy to be satisfied, but the performance requirement is more difficult to meet, and it takes nontrivial software expertise and skills to build a high-performance networking system. The packet forward mechanism such as run-to-completion (RTC) model, or pipeline model, is a good reference to build the real-world performant packet processing workloads on the multicore CPU architectures.

In DPDK, Packet Framework attempts to define the core libraries for building complex packet processing pipelines, thus providing reusable software code and extensible template. The commonly used pipeline blocks such as ACL (access control list) (for firewall), LPM (longest prefix matching) and exact matching, and hash lookup (for routing and flow classification) are built in DPDK. This chapter first examines the hardware-based forwarding model and then walks through the DPDK forwarding models and frequently used libraries and algorithms.

5.1 FORWARDING MODELS

Traditionally, a typical packet processing system consists of hardware/software co-design. Generally, a life cycle of packet goes through the system in the following order:

- *Packet input*: An incoming packet arrives at the hardware interface.
- *Preprocessing*: Parse the packet header in a coarse-grained manner.
- *Input classification*: Classify the packet in a fine-grained manner.
- *Ingress queuing*: Insert the packet into a FIFO (first in, first out) queue.
- *Delivery/scheduling*: Schedule based on queue priority and CPU status.
- *Accelerator*: Dedicated functions such as encryption/decryption and compression/decompression (e.g., IPsec).
- *Parallel CPU processing*: Complex processing via software, such as protocol and flow session management.
- *Egress queuing*: Schedule at packet exits according to QoS policy.
- *Post-processing*: Release the packet buffer in the post-processing.
- *Packet output*: Send the packet out through the hardware interface.

As shown in Figure 5.1, some functions (I/O processing) can be implemented by hardware-based modules (using NIC, FPGA card, or other accelerator). Based on the specific system needs, the system architect evaluates the hardware (chip) function, performance, cost, and quality, and selects the appropriate NIC and/or hareware accelerator. For the complicated tasks, software-based modules are preferred to run on the general-purpose computing processor, and these modules can be application, service packet header parser, and session management.

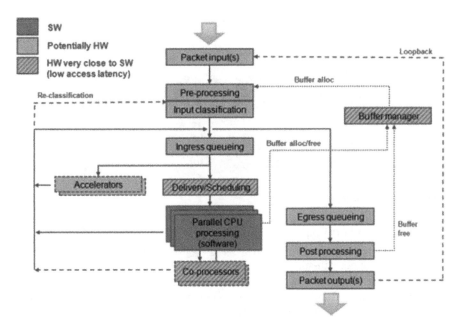

FIGURE 5.1 A general packet processing procedure.

In the early days, most of the packet processing functions were implemented by the purpose-built silicon. Usually, the chip can implement the pipeline model and/or RTC model, or even hybrid.

5.1.1 Pipeline Model

The pipeline model, as the name suggests, is inspired by the industrial pipelining assembly line model. It divides a complex function (above the modular level) into several individual stages, and each stage is connected via queues. Usually, the networking system includes many compute- and I/O-intensive tasks; the pipeline model allows us to execute the compute-intensive tasks on a microprocessing engine and the I/O-intensive operations on another microprocessing engine. Each processing engine is task centric. Data are shared using queues between the tasks. Queues should have enough bandwidth to match the speed of the processing so that they will not become the bottleneck. Also, the latency of the whole pipeline path is an important factor that should be considered carefully. In a pipeline model, the system throughput is bounded by the slowest processing stage. Ideally, each stage should have the same execution cost to maximize the efficiency (Figure 5.2).

In the below example, the most important component is the packet processing engines; there are many parallel execution units; every unit is a special microprocessing unit optimized for a specific task. A packet will go through the multiple task units after the input processing. The output from one task unit will become the input of the next task unit. The hardware pipeline determines the path of packet processing which is chained with the sequence of task units.

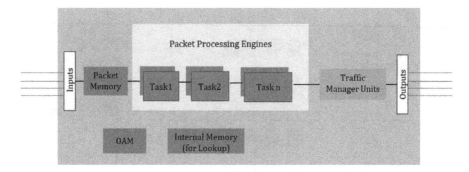

FIGURE 5.2   NPU pipeline model.

The hardware-based pipeline can be limited on its programmability; the hardware resource is decided during the chip design phase; each hardware execution unit is designed for a domain-specific task. Another limit is the hardware resource. For example, the on-chip memory can help the low-latency table lookup, but the on-chip memory size is often limited. A new hardware chip design may take 3–4 years from concept to the final product release. The development cadence is subject to business reason (return and investment).; the hardware networking infrastructure refresh cycle is usually slower than the server infrastructure.

5.1.2   RTC MODEL

The run-to-completion (RTC or known as "run to end") model is relatively simple to understand. A program has the execution engine (e.g a CPU core) ownership; it will run until the task is completed. RTC is heavily used in DPDK and software virtual switches. And in the higher level, the recent cloud native trend has a similar concept to RTC model; the cloud workload instance is spin up as needed, and shut down when the task is completed. If there are more service needed, one can just launch more workload instances to meet more service needs. Simply replicating the same tasks on more cores can gain scalable performance. The RTC model fits the charateristics of multicore system very well. A modern x86 processor provides tens of identical cores for general-purpose computing. Software can scale up easily across multiple cores. DPDK leverages this model, and using more core can easily improve the total packet throughput. RTC model doesn't necessarily share data via queues among cores, as the entire task is completed on its own execution resource.

RTC is a sequential execution model; the whole path can be a series of complicated tasks, so it may include many memory accesses. Each memory access will cause additional access latency. For each memory access, core has to wait for the completion of data load/store. The latency may accumulate so that the overall latency will increase.

RTC model is popular. It meets the silicon development trend. CPU is in a rapid product cadence to deliver more cores every 1–2 years. This makes the general-purpose server platform a very competitive candidate to deliver the software-based network function infrastructure. Software-based packet forwarding model has faster development cycle than hardware/software co-design model, and software provides high portability and low development cost. It aligns well with cloud-based elastic model. Which enables on-demand use of the hardware (core resources can be assigned dynamically for high or low network function needs). Telecom operators see this technology as an opportunity to reduce the capital investment on networking infrastructures.

5.1.3   Hybrid Model

High-performance networking system needs to support high-speed I/O. CPU frequency is hitting the physical ceiling. While I/O interface speed increases significantly without signs of slowing down. The high-speed I/O (on NIC) can be handled by RTC model; more cores can work together with NIC RSS (receive side scaling) and the use of multi-queue. This relieves the I/O bottleneck on the server platform. Cisco demonstrated 1TB route forwarding on Intel® server platform in 2017, which has been discussed in earlier chapter. In addition to packet I/O, network function system comprises the packet header parsing, protocol analysis, and packet modification; the whole processing takes many table lookup, which is translated into many times of memory accesses. The accumulated memory latency in RTC model is not an effective way to make good use of CPU.

- One way is to combine the multiple memory access together; this requires the software developer's mindset change; this is possible by merging the small table lookup into one large table lookup. But this approach requires code changes with the newly designed data structure. This may lead to less modular design due to performance trade-off, and lots of implementation changes.
- From the system architecture perspective, the other way to improve efficiency is to leverage the pipeline model; the pipeline model is good at hiding memory access latency. Break down the processing path into multiple tasks, then assign the different cores for the different tasks, and then connect all cores (tasks) to work together. The assigned core in the pipeline will only work on a specific task (as part of the pipeline stage), which improves the cache locality, thus resulting in higher efficiency. The major benefit of using the pipeline model is cache locality. Each core only executes a small portion of the whole application; thus, it is more friendly to the instruction cache. It is also easier for hardware and software prefetching to read data ahead of time. Additionally, the pipeline model breaks down the whole processing tasks into small pieces; thus, each piece (pipeline stage) can scale up and down independently.

Hybrid model is not limited to x86 platform; it is widely used in many network processors or SoC (system on chip)-based systems. Intel® Architecture (IA) platform has already been proven to be efficient to run DPDK-based networking applications with many industry adoptions. DPDK supports the pipeline model, the RTC models, and also the hybrid model. General purpose CPUs such as x86 can run any of the model flexibly, and it is also easy to run other software tasks such as system control, management, and data analytics. All these tasks make a complete intelligent system.

As an example, RTC model is illustrated in Figure 5.3a. Core 0 can be the control thread, which does not participate in the packet handling. Core 1 and Core 2 are the examples of RTC models; each core is responsible for the entire life cycle of packet processing, from Rx to Tx; Cores 3, 4, and 5 are using a similar RTC model; the difference is that NIC may turn on RSS features, so that the arrived packets may be processed by Cores 3, 4, and 5 for load balancing purpose. Pipeline model is illustrated in Figure 5.3b. The packet processing is divided into logical stages A, B, and C. A packet needs to go through Rx -> App A -> App B -> App C -> Tx stages. Rx/Tx is supported by Core 0. Depending on the workload complexity, both Core 1 and Core 2 are running the same tasks (App A -> App B ->App C); eventually, the packet is returned to Core 0 for packet Tx.

In general, the software pipeline has much more flexibility than the hardware pipeline. General-purpose core can do any tasks, but the HW-based task-optimized processing (TOP) units are by nature inextensible and limited to a specific domain. Networking technology is always innovating with new protocol, new data model, new algorithm, and new service. The performance advantage of the HW pipeline model has diminished in the recent years. x86 processor is becoming a cost-effective choice because software is catching up with the performance goal.

Both the pipeline and RTC models are easy to be developed. The pipeline model is good at latency hiding. In order to pursue more performance, sophisticated workload-specific performance tuning is required. RTC model is much simpler; it depends on the high core counts to scale up the performance. Hybrid use is also common in complicated systems.

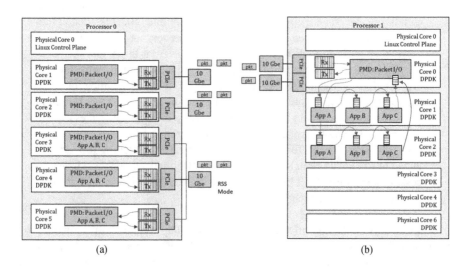

FIGURE 5.3   Forwarding models. (a) RTC model and RSS and, (b) pipeline model.

TABLE 5.1
Comparing RTC and Pipeline Models

Perspective	Pipeline Model	RTC Model
Development	Easy, x86 core is general purpose, can do any task. Packet Framework is an existing implementation; the user can use it as a reference to build the features with software reuse (such as core/stages/tables, inter-core communication with queues, memory). Architecture focuses on the stage definition.	Easy. System consists of a list of function calls, often run as endless loop as a thread. Replicate to many cores.
Performance	Good way to hide the memory latency, I/O, and load balancing. Add the overhead between cores, workload-specific tuning on the pipeline stages.	Replicate to many cores. Increasing cache locality is a challenge. Reducing memory-related processing can be achieved with code refactoring and design optimization.

5.2   DPDK SPECIFICS

5.2.1   RTC MODEL

With network driver in Linux kernel, an interrupt happens with the packet arrival, and core is waken up to handle the packet with interrupt service routine. The actual core to service interrupt is decided by OS, and the load balancing among cores is implemented by OS as well. Linux kernel driver can use more cores to achieve high performance in the scenario of large amount of incoming traffic.

DPDK initialization will reserve the specific cores to work on the NIC ports for packet Rx/Tx. If there are more NIC ports, a software developer can configure more cores. For the high-speed interface, the software developer can configure multiple queues for each interface, then assign multiple cores to do packet processing for those queues. This is a common approach of DPDK to realize performance scaling. With RTC model, the entire life cycle of a packet (illustrated earlier) can still be handled within one thread, starting from packet Rx, classification, filtering, QoS, and scheduling, until the final step of packet Tx. In 2019, Intel® Xeon processor supports up to 56 physical cores (112 logical cores); each core has security-related instruction sets such as AES-NI and SHA-NI, which are very useful for data encryption/decryption and authentication. Most of DPDK sample applications demonstrate the use of the poll-mode driver, which means the core works in a continuous loop to poll packets from the interface. Although the poll-mode ensures each incoming packet to be processed as soon as possible, the CPU core runs at 100% all the time which may lead to higher power consumpion. Developers should consider interrupt mode if the packets are coming sporadically. Once the packet enters the system, the packet processing will continue in software. Packet Tx is often the last step to complete the processing task. When the traffic does not come in high burst at the line rate, system designers need to consider the other design model, including the event-based mechanism (like interrupt).

When large traffic floods into the system, cache size is always not big enough, and it is likely that more cache misses will happen. Simply speaking, assigning more cores to replicate the packet processing task on more instances can reduce the burden on each working core. When new cores are assigned for RTC model, it is natural to expect the traffic is evenly distributed to cores. NIC acts as the ingress device, and it is recommended to be configured with multi-queues together with flow classification offload (such as RSS, or other flow classification). NIC can look at the incoming traffic and direct them to different ports/queues, where each service core can handle packets with RTC model to be scalable.

Internet traffic has the tidal effect, and there are high traffic time and low traffic time. A smart way is needed to assign more or less cores to handle the high or low traffic scenario. It is beyond the scope of DPDK. NIC has statistics that gather packet arrival rates; the system can collect and monitor the traffic load in the runtime to make an intelligent decision; and it can be configured dynamically with more queues, so that more cores can be assigned to take care of the incoming packets on these interfaces.

5.2.2 PIPELINE MODEL

DPDK pipeline applications can be easily developed using DPDK Packet Framework, which offers a modular and flexible framework that leverages highly optimized packet I/O libraries and low-level processing primitives for high performance. The key design objectives are listed as follows:

- Provides standard methodology to build complex packet processing pipelines.
- Provides reusable and extensible templates for the commonly used applications' functional blocks (pipeline stages).
- Provides capability to switch between pure software- and hardware-accelerated implementations for the same functional block.
- Provides implementation flexibility in terms of selecting the programming model such as RTC, pipeline model, or hybrid model on multicore CPUs to achieve the best trade-off between flexibility and performance.
- Provides a framework that is logically similar to OpenFlow model that comprises a pipeline of match-action tables (MATs).

The Packet Framework leverages OpenFlow design principles (port, table, pipeline, actions, etc.) to build a reference model. It introduces three new DPDK libraries:

- librte_port;
- librte_table;
- librte_pipeline.

The aforementioned libraries are helpful in defining a reusable and extensible template for pipeline modules. Each pipeline module has three components: input port, output port, and table, as shown in Figure 5.4. The pipeline is constructed by connecting the set of input ports with the set of output ports through a chain of tables. Input ports can only be connected to one table. However, tables can be connected to other tables and also to the output ports (Figure 5.4).

In the multicore environment, it is possible to have multiple pipelines running on different cores, as shown in Figure 5.5. The pipelines are connected together with the purpose of passing the packets from one block to the next blocks, so that the application functionality gets implemented. This way, a topology of interconnected

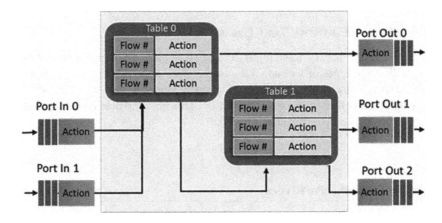

FIGURE 5.4   DPDK Packet Framework: pipeline block.

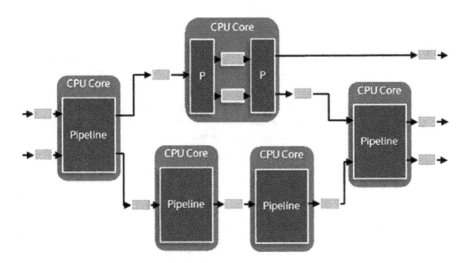

FIGURE 5.5   Multicore for application pipeline.

pipelines gets defined, as well as a topology of interconnected cores, e.g., pipeline, RTC, and hybrid. Thus, the framework allows application developers to provide enough flexibility in mapping a set of pipeline blocks to physical cores to meet performance requirement (Figure 5.5).

5.2.3 PACKET FRAMEWORK PORT LIBRARY

The port library (librte _ port) provides a standard interface to implement different types of packet ports. This library defines initialization and runtime operation of the packet ports. Table 5.2 shows the list of DPDK entities abstracted as input/output ports using the port library (Table 5.2).

5.2.4 PACKET FRAMEWORK TABLE LIBRARY

Similar to the port library, table library (librte _ table) offers a standard interface for different types of lookup tables used in data plane processing, as shown in Table 5.3. Any search algorithm that can uniquely associate data with a lookup key can be fitted under the lookup table abstraction. This library defines the initialization and runtime operation of the table (Table 5.3).

TABLE 5.2
Packet Framework—Port Types

#	Port Type	Description
1	SW Ring	A software circular buffer for packet/ message transfer between pipeline modules.
2	HW Ring	Queue of buffer descriptors used to interact with NIC, switch, or accelerator ports.
3	IP Reassembly	Assembles IP fragments to an IP datagram.
4	IP Fragmentation	Fragments input packets with IP datagrams larger than MTU (maximum transmission unit) size.
5	Traffic Manager	Performs congestion management and hierarchical scheduling as per service-level agreements on an egress port.
6	Source Port	Packet generator port, similar to Linux kernel/dev/zero character device.
7	Sink Port	Packet terminator port to drop all input packets, similar to Linux kernel /dev/null character device.
8	Kernel NIC Interface (KNI) Port	Interface for sending/receiving packets to/ from Linux kernel space.

TABLE 5.3

Packet Framework—Table Types

#	Table	Description
1	Hash Table	Exact match table; the lookup key is an n-tuple of packet fields that uniquely identifies a traffic flow. It is used to implement flow classification tables, ARP tables, etc.
2	Longest Prefix Match (LPM)	The lookup key is an IPv4/IPv6 address. The lookup operation selects the longest IP prefix that matches the lookup IP address. Typically used to implement layer 3 routing tables.
3	Access Control List (ACL)	The lookup key is n-tuples of packet fields. The lookup operation searches over a set of rules and finds the best match (highest priority). It is used to implement rule base for firewalls, etc.
4	Array Table	The lookup key is a table entry index. It is used for implementing small databases such as IPv4 DSCP field entries, etc.

5.2.5 PACKET FRAMEWORK PIPELINE LIBRARY

The pipeline library (librte _ pipeline) employs performance-optimized "librte _ port" and "librte _ table" to construct the application functional blocks (pipelines). Once all the elements (input ports, tables, output ports) of the pipeline module are created, the input ports are connected to the tables, table action handlers are configured, and the tables are populated with the set of entries and associated actions. For each table, behaviors of user actions for lookup hit and miss cases are defined using the table action handlers, whereas the next hop action for packets (either another table, an output port, or packet drop) is determined through reserved actions. In order to speed up DPDK application development, the pipeline libraries provide a common set of actions for the pipeline tables. The pipeline tables can be updated during runtime, and the input ports can be disabled or enabled later on as required.

Figure 5.6 shows the packet processing flow in the pipeline module. When a pipeline runs, it scans through all the input ports and fetches the available packets (a maximum of 64 packets are processed at a time). If any input port has an associated action, it will be performed on all the packets entering through the port. Afterward, the packets are pushed towards the table connected to that input port. The result of the table lookup operation could be a hit or miss.

- In case of lookup miss, an associated user-defined action, if configured, is executed and the default entry action (packet drop or send to the output port) is applied.

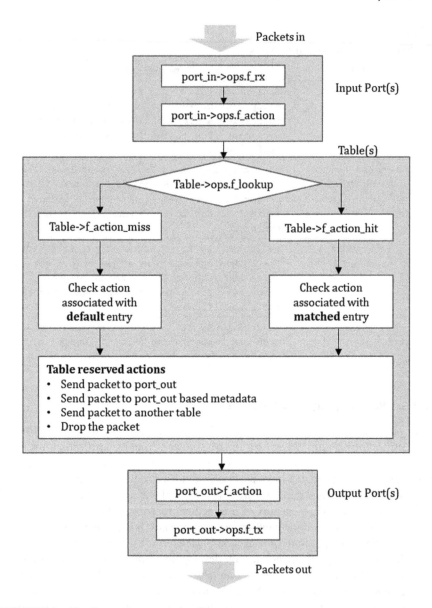

FIGURE 5.6   Pipeline packet processing flow.

- Likewise, in case of lookup hit, similar steps are performed, and finally, a matched entry action (one of the reserved actions) is executed to decide the next hop for the packet.

All the packets directed towards the output port are first subjected to port action handlers and then sent out of the pipeline output port (Figure 5.6).

5.3   FORWARDING TABLE AND ALGORITHM

In addition to the framework, a very important part in the forwarding path is to match and classify the packet header. The flow matching and classification is done against the packet forwarding tables. DPDK has the specific optimized implementations such as Hash libs, LPM libs, ACL, and Event tables. The classification algorithms can be categorized as follows:

- *Exact match algorithms*: Parse the packet header to exactly match all the bits in a pattern before the next action.
- *Wildcard match algorithms*: They are more relaxed than an exact match. They match partial header with wildcard patterns (e.g., LPM, ACL lookup in DPDK).

5.3.1   HASH TABLE

Hash table is the most common data structure used for the exact match. It is widely used in networking functions such as flow tables for connection tracking and NAT (network address translation) table. Ideally, hash table has O(1) lookup time. Insertion and deletion times are also important for evaluating a hash table design since they are closely tied to the performance. Another critical parameter commonly used for describing a hash table is the load factor (L). Load factor describes how full a hash table is. With different hash table designs and table lookup, insertion and deletion times may vary depending on the load factor.

$$L = \frac{n}{k}$$

n = Number of keys stored in the hash table;
k = Total entries of the hash table.

During key insertion into a hash table, keys are hashed and placed into the hash table at specific locations; according to the hash value, hash collision could happen. When two different keys have the same hash value and collide, the latter key might be failed to be inserted into the table. Hash tables use different algorithms to handle the hash collision. In this chapter, we focus on the default hash table algorithm in DPDK (hash library), which is based on cuckoo hashing algorithm.

5.3.2   HASH LIBRARY

Cuckoo hash was described in Rasmus Pagh and Flemming Friche Rodler's paper "Cuckoo Hashing", and DPDK's hash table implementation is an extension based on cuckoo hashing. The hash table is implemented as two tables. The first table is bucket-based signature table. It consists of an array of buckets, and each bucket has a fixed number of entries. Each entry contains a short signature of a key and an index to the key data pair. The second table consists of an array of key data pairs. When looking up a key, the signature in the first table will be compared first, and the key data pair in the second table is accessed only if the signature matches.

Each key can be stored in one of the two candidate buckets in the hash table. During insertion, the cuckoo hashing algorithm is used to make space for the new key to insert if the primary bucket of the key is full. For example, if the newly inserted key A found its primary bucket is full, then key A will kick key B from the bucket to key B's alternative bucket. This process is recursive, meaning key B could kick key C to key C's alternative bucket. This cuckoo displacement continues until every key finds its new location and key A is successfully inserted. With the cuckoo algorithm, DPDK hash library can achieve up to 95% of the load factor before the first key insertion failure happens.

For lookup, the key will be hashed to locate both candidate buckets and to have a short signature. The entries of each bucket will be iterated, and the signature of each entry will be compared against key's signature. If the signature matches, the key in the second table will be compared against the lookup key. Only if both signature and key match, the resulting data will be retrieved.

The algorithms described above ensure the DPDK hash table to have a fast O(1) lookup time and high load factor before the first insertion failure. For example, the signature table avoids unnecessary key comparisons. This is very useful in the case of long key size. The bucket-based data structure and the cuckoo algorithm ensure that the key insertion should not fail easily due to hash collision.

Besides the high-level algorithm and data structure design, the hash library is also enhanced by various low-level software techniques to improve its speed. For example, SIMD instructions like Intel® Advanced Vector Extensions (Intel® AVX) are used to compare multiple signatures together. Intel® Transactional Synchronization Extensions (Intel® TSX) instructions are used to realize thread safety. Software prefetches and key batching are also used across the code to minimize the cache access time. For more information, please refer to the DPDK source code and programmer's guide.

5.3.3 Elastic Flow Distributor

Elastic flow distributor (EFD) library is another DPDK library providing an exact matching feature. The purpose of this library is to provide an efficient data structure in order to help distribute flows to specific targets (e.g., backend machines). EFD is based on the research work from *D. Zhou. et al., Scaling Up Clustered Network Appliances with ScaleBricks.*

Conceptually, one can easily create a flow distributor by a simple hash function. For example, if there are eight targets in total, by hashing the flow key and mod eight, the flow will be directed to a specified target accordingly and never goes to another target. A more complex algorithm can be used. For example, the consistent hashing algorithm is usually used to handle dynamic change in the number of targets.

However, only using a hash function (computation based) to distribute flow does not give the operator the capability to fully control which target each flow should go to. The target is only decided by the predefined hash function. To have the capability to specify the target for each flow, one straightforward approach is to use a hash table to store the key-target mapping. For example, the operator can insert the flow and its specified target into the hash table. Later on, a hash table lookup can find the target for this flow.

One drawback of using the hash table is the large storage requirement. Considering that the distributor needs to handle millions of flows, then the hash table needs at least millions of entries to save all the information. EFD comes to solve this issue.

The fundamental idea of EFD is to find the perfect hash functions to map each key to its corresponding target. Thus, the key does not need to be compared to lookup the mapping. The hash functions themselves are sufficient to calculate the mapping. To illustrate the idea, the following simple example can be used for the table construction. If we want to do the below,

$$\text{"Key a"} -> \text{Target 3} (0b11)$$

$$\text{"Key b"} -> \text{Target 1} (0b01)$$

$$\text{"Key c"} -> \text{Target 0} (0b00)$$

In a hash table approach, one just needs to store keys a, b, and c and their targets into the hash table. However, if the key length is long, the hash table requires a lot of memory access during lookup. For EFD, it stores two hash functions h1 and h2. During insertion, from a hash function pool, EFD picks up two hash functions that satisfy the following condition:

$$h1(a) = 1, h1(b) = 0, \text{ and } h1(c) = 0,$$

$$h2(a) = 1, h2(b) = 1, \text{ and } h2(c) = 0.$$

Now, to lookup "Key a", let's first calculate the following:

$$h1(a) \text{ will get the first bit} = 1.$$

$$h2(a) \text{ will get the second bit} = 0.$$

Thus, by just calculating these two hash functions, it avoids lots of memory access to compare "Key a" with a hash table entry, and key's target is easy to get.

$$\text{Key} -> \text{target}$$

$$a -> 11$$

$$b -> 01$$

$$c -> 00$$

In the real implementation, hash functions are represented by an index into the hash function pool, so in this case we only read two index numbers during lookup, which do not use much cache memory. To be easier to find the perfect hash functions, keys are assembled into small groups, so the perfect hash functions are searched within a small group of keys.

Another library that provides a similar feature is the membership library. DPDK membership library provides the feature to test whether a key is a member of a given data set or not. Bloom filter is one of the classic filter data structures. In a bloom filter, each key is hashed by multiple hash functions to index a bitmap. If all of the indexed locations of the bitmap are set to 1, then the key belongs to this data set. During insertion, a new key will set all the corresponding locations of the bitmap to 1. As one may imagine, bloom filter does not provide 100% accuracy. False positives are allowed for filter data structure. In other words, if a key is found to belong to the data set, it has a small possibility that this is not the case. However, similar to EFD, the bloom filter provides a compact data structure compared to the hash table. Due to the property of a filter data structure, it suites for many networking use cases that require a fast lookup, while the false positive can be tolerated. Some use cases include flow caching, load balancer, and routing loop prevention.

The membership library in DPDK provides a couple of different filter data structures; each data structure has a slightly different design and property. For example, a vector of bloom filters is provided to serve as an array of bloom filters. With this data structure, one can check if a key belongs to a certain data set of all the data sets represented by the multiple bloom filters. Another filter data structure is hash table-based filter, which is based on cuckoo filter (*B. Fan, et al, Cuckoo Filter: Practically Better Than Bloom.*). Compared to traditional hash table, cuckoo filter does not store the full key but only a signature of the key into the table. Without the full key comparison, there could be false positives while the memory requirement is much less.

EFD and membership library have a similar purpose in the sense that both data structures try to use much less memory but do a similar job as a hash table. They provide users more options where a hash table usually applies. Generally speaking, both EFD and filters should have faster lookup speed than the hash table. Meanwhile, EFD may have slower insertion rate because of the perfect hash searching process, and filters usually have false positives. It depends on the use case to choose the best algorithm.

5.3.4 Longest Prefix Matching

LPM is a typical routing table lookup algorithm. Every entry in the LPM table specifies a subnet, and a destination address may match with multiple entries in the table. The one with the longest prefix (subnet mask) is called the longest prefix match and also as the winning entry. DPDK designed an LPM lookup table, which achieves high performance at the cost of large memory consumption. LPM's main data structure includes the following:

- *tlb24*: A table with 2^{24} entries;
- *tlb8*: Multiple tables with 2^8 entries.

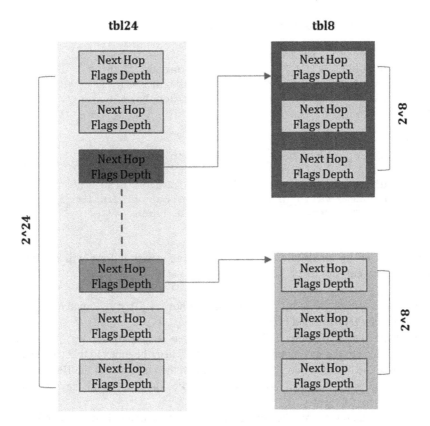

FIGURE 5.7   LPM diagram.

It is obvious that 24-bit prefix may have a total of 2^24 entries. Each entry corresponds to 24-bit prefix, which is associated with the last 8-bit suffix. The last 256 entries (8 bits) can be assigned on demand so that both space and time are taken into account.

- When the prefix length of the lookup table entry is less than 24 bits, it is able to find the next hop with one memory access;
- When the prefix length is more than 24 bits, it will become two memory accesses.

This design is based on the known probability and statistics study; the probability of the former case is higher, while that of the latter case is lower. Unlike the traditional embedded system, large memory is present in the server system (Figure 5.7).

```
struct rte_lpm_tbl24_entry {
        /* Stores Next hop or group index (i.e. gindex)into
        tbl8. */
                union {
                uint8_t next_hop;
```

```
          uint8_t tbl8_gindex;
      };
      /* Using single uint8_t to store 3 values. */
      uint8_t valid    :1; /**< Validation flag. */
      uint8_t ext_entry :1; /**< External entry. */
      uint8_t depth    :6; /**< Rule depth. */
};

struct rte_lpm_tbl8_entry {
      uint8_t next_hop; /**< next hop. */
      /* Using single uint8_t to store 3 values. */
      uint8_t valid    :1; /**< Validation flag. */
      uint8_t valid_group :1; /**< Group validation flag. */
      uint8_t depth    :6; /**< Rule depth. */
};
```

When looking up on IP address, the first 24 bits of IP address will be matched against every entry of tbl24.

- If the "valid" bit is set and ext_entry is not set (0), it is a match and the ext_hop tells the next hop.
- If the "valid" bit is set and ext_entry is also set (not 0), the next_hop field will lead to the second table lookup (tbl8). Then, the index to tbl8 can determine the matched entry, which is based on the last 8 bits of the IP address. Furthermore, the next hop address can be found from the next_hop of rte_lpm_tbl8_entry.
- If no match is found, the default next hop will be taken. User can configure the next hop.

To evaluate the lookup performance, DPDK provides a unit test code for the actual measurement on any specific platform. Please refer to the DPDK source code to get more information.

5.3.5   ACL

ACL (Access Control List) is another classical packet-based security function; the incoming packet is parsed with "n-tuple"; the packet-specific n-tuple is gathered as "matching fields". It will match against the ACL rules. If matching succeeds, the action can be "Drop (Deny)", "Forward (Permit)", etc. DPDK provides a software implementation, and ACL usage is listed below (Table 5.4):

- Create an access control context.
- Add rules to the AC context.
- Build the runtime data for all the defined rules.
- Perform rule-based matching on the incoming packets.
- Destroy AC context.

TABLE 5.4

ACL APIs

ACL API	Function Description
rte_acl_create	Create an AC context.
rte_acl_add_rules	Add rules to the AC context.
rte_acl_build	Create a structure for AC runtime matching.
rte_acl_classify	Match the AC rule.

The ACL library in DPDK is optimized with SIMD instructions, which allow that the match can happen to multiple packets; the software algorithm can take advantage of SSE (streaming SIMD extensions) and AVX2 at the runtime. One limitation of ACL library is that it requires rebuilding of ACL contexts whenever there are rule updates.

Here is a key data structure on matching fields:

```
struct rte_acl_field_def {
        uint8_t type;      /**< type - RTE_ACL_FIELD_TYPE_*. */
        uint8_t size;      /**< size of field 1,2,4, or 8. */
        uint8_t field_index; /**< index of field inside the
        rule. */
        uint8_t input_index; /**< 0-N input index. */
        uint32_t offset;   /**< offset to start of field. */
};
```

For IPv4-based packet matching, 5-tuple example is used here, which consists of protocol field, source IP address, destination IP address, source port, and destination port. Here is the configuration example.

```
struct rte_acl_field_def ipv4_defs[NUM_FIELDS_IPV4] = {
    {
                .type = RTE_ACL_FIELD_TYPE_BITMASK,
                .size = sizeof(uint8_t),
                .field_index = PROTO_FIELD_IPV4,
                .input_index = PROTO_FIELD_IPV4,
                .offset = offsetof(struct ipv4_5tuple, proto),
    },
    {
                .type = RTE_ACL_FIELD_TYPE_MASK,
                .size = sizeof(uint32_t),
                .field_index = SRC_FIELD_IPV4,
                .input_index = SRC_FIELD_IPV4,
                .offset = offsetof(struct ipv4_5tuple, ip_src),
    },
    {
                .type = RTE_ACL_FIELD_TYPE_MASK,
                .size = sizeof(uint32_t),
```

```
                .field_index = DST_FIELD_IPV4,
                .input_index = DST_FIELD_IPV4,
                .offset = offsetof(struct ipv4_5tuple, ip _ dst),
       },
       {

                .type = RTE_ACL_FIELD_TYPE_RANGE,
                .size = sizeof(uint16_t),
                .field_index = SRCP_FIELD_IPV4,
                .input_index = SRCP_FIELD_IPV4,
                .offset = offsetof(struct ipv4_5tuple, port _ src),
       },
       {

                .type = RTE_ACL_FIELD_TYPE_RANGE,
                .size = sizeof(uint16_t),
                .field_index = DSTP_FIELD_IPV4,
                .input_index = SRCP_FIELD_IPV4,
                .offset = offsetof(struct ipv4_5tuple, port _ dst),
       },
         };
```

5.4 EVENT SCHEDULING

Event/packet scheduling focuses on addressing how the packets are scheduled from I/O device to core, from core to core, and from core to device. The cost of event scheduling could be high if implemented in software. Hardware accelerator can help in some cases.

DPDK supports a few event/packet scheduling approach:

- NIC-based packet scheduler;
- Software-based packet distributor;
- Event device framework.

The packet flow can be defined in various ways. One common way is based on 5 tuples (src_ip, dest_ip, src_port, dest_port, and tos). Different 5-tuples can be classified as different packet flows for a network middle box. There are three packet-based event scheduling types:

- *Parallel*: Packets from the same flow can be distributed to multiple cores at the same time, without packet ordering requirement.
- *Ordered*: Packets from the same flow can be distributed to multiple cores but the packet order has to be restored at the egress port to the same order as ingress.
- *Atomic*: Only one packet from the same flow is processed at one time, system wide.

There are multiple event sources in a system. Packet can be one event source, and the time-based event is another kind. This chapter focuses on the packet-based event.

5.4.1 NIC-Based Packet Scheduler

RSS and FDir are NIC-based technologies, which will be introduced in the later chapter. RSS distributes the received packets into multi-queues for load balancing among multicores. FDir directs the packet flow to the specific queue, which ensures the target core is the same one that handles the entire packet flow. The exact approach is NIC specific. Both RSS and FDir may not handle the complex packet header parsing like looking into the multiple layers of tunneling protocol.

5.4.2 Software-Based Packet Distributor

When the packet is received by a CPU core, it may be sent to other cores for further packet processing. After processing, the packets from the same flow are usually expected to be sent out in the same order as they arrived.

For example, in a pipeline model, there are four packets belonging to the same flow arriving at the Rx core. Then, Packets 1 and 3 are scheduled to Cores 1 and 3, and Packets 2 and 4 are scheduled to Cores 2 and 4, respectively. After processing, the original order of the packets may be totally lost if there is no reordering mechanism (Figure 5.8).

Therefore, the software packet scheduling has two key problems:

- How to maintain the packet ordering in a parallel system.
- How to achieve it with high performance.

DPDK provides an example, **packet distributor**, to demonstrate how to distribute packets among cores efficiently; the example code is located at /dpdk/examples/distributor/.

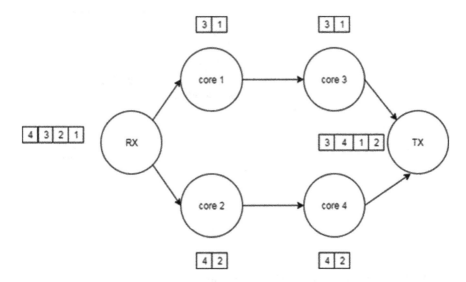

FIGURE 5.8 Packet distributor example.

The distributor example distributes the packets from an Rx port to multiple cores. When the packet is processed by the core, the destination port of a packet is the port from the enabled port mask adjacent to the one on which the packet was received. The distributor example consists of four types of threads: a RX thread, a Distributor Thread, a set of Worker Threads, and a TX thread. How these threads work together is shown in Figure 5.9.

The "RX thread" receives the packets and enqueues them to an rte_ring. The "Distributor Thread" dequeues the packets from the ring and assigns them to "Worker Threads". This assignment is based on the tag (or flow ID) of the packet. The "Distributor Thread" communicates with the "Worker Threads" using a cache line swapping mechanism, passing up to 8 mbuf pointers at a time (one cache line) for each worker.

More than one "Worker Thread" can exist, and these "Worker Threads" do packet processing by requesting the packets from the distributor, carrying out a simple XOR operation on the input port (to find the output port which will be used later for packet transmission), and then returning the packets to the "Distributor Thread".

The "Distributor Thread" will get the processed packets and enqueue them to another rte_ring. The "TX thread" will dequeue the packets from the ring and transmit them to the output port, which is specified in packet mbuf (rte_mbuf).

The key design for the "Distributor Thread" is the data structure which is exchanged among cores. Inter-core data path is a known bottleneck of packet scheduling. The data structure needs to be cache-friendly because the cores talk with each other through last level cache (LLC). For x86, each cache line size is 64 bytes.

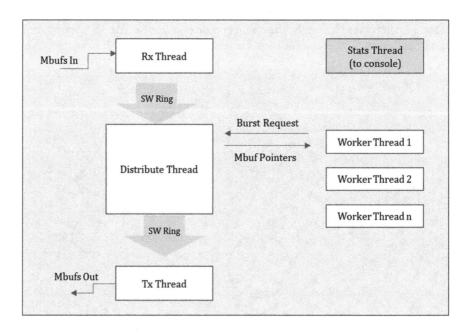

FIGURE 5.9   Packet distributor implementation flow.

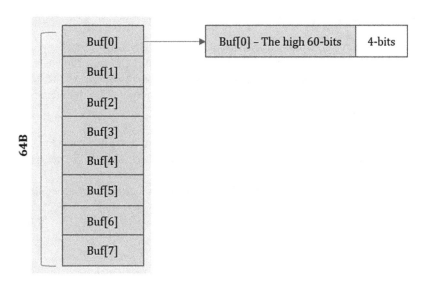

FIGURE 5.10 Packet distributor—packet buffer array.

Each cache line can store up to eight pointers, which is 64 bits wide. The data structure being used to exchange packets is illustrated in Figure 5.10.

There is an assumption that all mbuf address is 64 bytes aligned, which means the least significant 4 bits is always 0. Thus, the least significant 4 bits can be used for other purposes, such as mbuf status flag in the following:

```
#define RTE_DISTRIB_NO_BUF 0 /**< empty flags: no buffer
requested */
#define RTE_DISTRIB_GET_BUF (1)/**< worker requests a buffer,
returns old */
#define RTE_DISTRIB_RETURN_BUF (2)/**< worker returns a
buffer, no request */
```

Ideally, the "Distributor Thread" can send 8 mbuf pointers to the worker core every time, and the worker core returns 8 mbuf pointers after completing the batched packet processing. This cache-aligned communication between "Distributor" and "Worker" threads minimizes the cache coherence overhead.

The packet distributor application does not address the packet reordering issue. In addition, two other things to consider when developing a packet distributor based on these sample applications:

- Distributor needs a dedicated core for the packet scheduling work.
- The Worker Thread cannot handle the asynchronous task. The dispatching workflow is based on a blocking model, which cannot track the asynchronous packet.

5.4.3 EVENT DEVICE

In order to achieve high performance for more complicated scheduling tasks (atomic, ordered), DPDK introduces the event-driven programming model, known as "eventdev". The eventdev library simplifies the packet processing with automatic multicore scaling, dynamic load balancing, pipelining, packet ingress order maintenance, and synchronization services.

- In a polling model, core polls NIC ports/queues to receive a packet.
- In an event-driven model, core polls the scheduler to select a packet.

DPDK supports HW-based event scheduler and SW-based event scheduler. One example is located at dpdk/examples.eventdev _ pipeline. For scheduler, regardless of hardware or software implementation, it includes the following attributes:

- *Flow_id*: Indicates which flow this event/packet belongs to.
- *Priority*: Indicates the event priority level from high to low.
- *Sched_type*: Represents the type of scheduling that should be performed on this event, and valid values include ORDERED, ATOMIC, and PARALLEL.
- Queue_id: The identifier for the event queue that the event is sent to.
- *Port*: The points of contact between worker cores and the eventdev.

The eventdev is composed of port, queue, and scheduler. Port can be enqueue port and dequeue port. After the event is populated, core can inject the event to the enqueue port. The injected event will be appended to the end of corresponding queue. The queue needs to be linked to certain dequeue port (1:1 or 1:n mapping). The scheduler checks each queue's status periodically; if the event is available to be dispatched, the scheduler will dispatch the event according to its configuration (e.g., schedule type, queue link) to the destination port. Eventually, core can poll the dequeue port and get the event which is dispatched there (Figure 5.11).

Let's look at an example life cycle of a packet here.

1. NIC receives the packet; *RX core* populates the flow id and then enqueues the packet to *Queue 0*.
2. *The scheduler* checks the event flow id and queue id and then dispatches the packet event to *Queue0::flow0* subqueue.
3. Core 0 polls event from the *Port 0* and then forwards this event to *Queue 1*, populates the queue id, and then enqueues event through *Port 0*.
4. *The scheduler* checks the event flow id and queue id and then delivers event to *Queue1::flow0* subqueue.
5. Core 1 polls event from *Port 1*, updates the event type as TX, and then enqueues the event by *Port 1*.
6. *The scheduler* checks the event type and then delivers the event to TX port.
7. *TX core* dequeues the event and transmits the packet out (Figure 5.12).

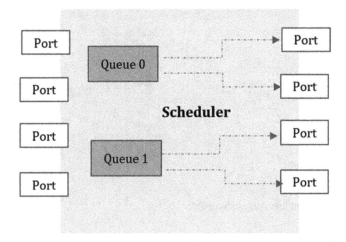

FIGURE 5.11 Event dev conceptual framework.

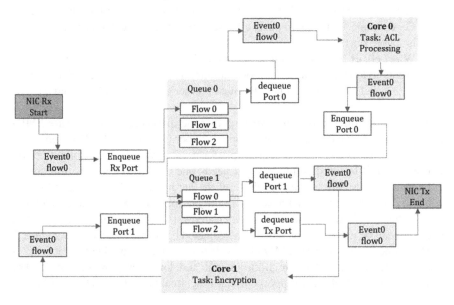

FIGURE 5.12 Event dev example in details.

We use examples to explain two scheduling types: atom and ordered

Atom Scheduler Type (Figure 5.13)
- If there are four events that belong to flow0.
- Scheduler only dispatches one event of flow0 at one time. Before the dispatched event enqueues again, no other event that belongs to flow0 will be dispatched.
- If event0 is enqueued, then event1 will be dispatched.
- And so on.

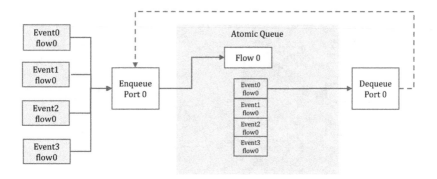

FIGURE 5.13   Eventdev: Atomic Scheduler

Ordered Schedule Type (Figure 5.14)
- There are four events that belong to flow 0.
- Scheduler dispatches e0, e1, e2, and e3 to different ports at the same time. For example, e3 is sent to por0, e0 is sent to port, e1 is sent to port2, and e2 is sent to port3.
- When e0, e1, e2, and e3 enqueue again, the scheduler will reorder all the events to keep them in order. For example, if the new enqueue order is e3, e1, e2, and e0, then the scheduler must reorder them as e0, e1, e2, and e3.

So far, there are three hardware PMDs (poll mode driver) and two software PMDs that support the eventdev. The application developer chooses the PMD according to the actual business use case on a given platform. Hardware PMD has good performance. The full-featured software PMD needs one dedicated core; the throughput and latency are usually worse than those of the hardware PMD. It is a good option if there is no hardware offload available in the platform. Developer can develop/test their applications based on the event-driven framework with software PMD first, then a decision can be made later on whether a dedicated hardware accelerator is required. OPDL is one of the software PMD library. It is particularly suitable for packet processing workloads that have high-throughput and low-latency requirements. The drawback is that OPDL only supports partial functionality of the eventdev APIs.

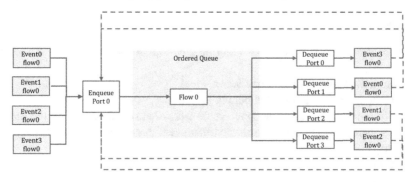

FIGURE 5.14   Eventdev: Ordered Schedule

5.5   IP_PIPELINE

The ip_pipeline is an example of DPDK, and the source code is located at /dpdk/ examples/ip _ pipeline/.

This application can create functional blocks (pipelines) based on the DPDK components such as performance-optimized input/output ports, tables, and actions in a modular way. Multiple pipelines can be interconnected through packet queues to create more feature-rich applications (superpipelines). The pipelines are mapped to threads, each pipeline can be executed by a single thread, and each thread can run one or several pipelines.

The pipeline creation is based on ports, tables, and actions, connecting multiple pipelines together and mapping the pipelines to execution threads. There are many possibilities, this is application specific, and system analysis and tuning might be required. The application can be built and managed through command line interface (CLI), which is flexible in development and performance optimization phases. CLI supports the following features in DPDK:

- *Primitive objects to create pipeline ports*: Memory pools, links (i.e., net-work interfaces), SW queues, traffic managers, etc.
- *Action profiles*: Used to define the actions to be executed by pipeline input/ output ports and tables.
- *Pipeline components*: Input/output ports, tables, pipelines, mapping of pipelines to execution threads (Figure 5.15).

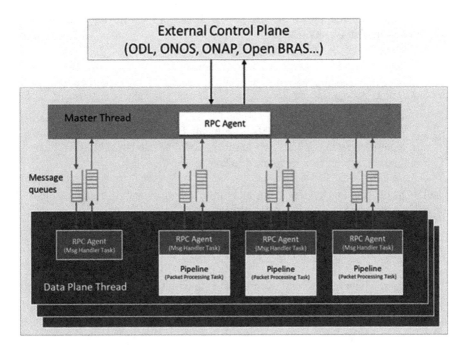

FIGURE 5.15   DPDK IP pipeline application.

An SDN (software defined networking) controller agent (e.g., ODL, ONOS, ONAP) or TCP client (e.g., telnet, netcat, custom script) can connect to the system, so the controller can issue commands through the network. System consists of master and data plane threads on different execution cores.

Master thread: It creates and manages the system, it is responsible for working with SDN controller, and it can accept the remote configuration via CLI input.

Data plane thread: Each data plane thread supports one or several pipelines. Each data plane thread can run the below tasks in time-sharing mode.

- *Packet processing*: Processes input packets from the input ports.
- *Message handling*: Periodically pauses the packet processing and handles the messages from the master thread such as:
  - Add/remove pipeline to/from current data plane thread.
  - Add/delete rules to/from the given table of a specific pipeline owned by the current data plane thread, read statistics, etc.

Start the ip_pipeline application. The startup command line is:

```
$ ip_pipeline [EAL_ARGS] -- [-s SCRIPT_FILE] [-h HOST] [-p
PORT]
```

The startup arguments are as follows:

- -s SCRIPT_FILE
  Optional: Yes, Default: Not present, Argument: Path to the CLI script file to be run at application startup. No CLI script file will run at startup if this argument is not present.
- -h HOST
  Optional: Yes, Default: 0.0.0.0, Argument: IP Address of the host running ip pipeline application to be used by remote TCP-based client (telnet, netcat, etc.) for connection.
- -p PORT
  Optional: Yes, Default: 8086, Argument: TCP port number at which the ip pipeline is running. This port number should be used by remote TCP client (telnet, netcat, etc.) to connect to host application.

The DPDK Getting Started Guide has more information on running applications and the environment abstraction layer (EAL) options.

To run ip_pipeline with a configured script (such as l2fwd):

```
$ ./ip_pipeline -c 0x3 -- -s examples/l2fwd.cli
```

To run remote client (e.g., telnet) to communicate with the ip_pipeline application:

```
$ telnet 127.0.0.1 8086
```

When both ip_pipeline and telnet client are running, telnet client can issue the command messages to ip_pipeline. At any stage, telnet client can be terminated using the quit command.

Several CLI examples are provided with ip_pipeline, and some examples are summarized in Table 5.5.

TABLE 5.5
IP Pipeline—CLI Examples

#	CLI Profile Name	CLI Messages
1.	L2Fwd (Table: Stub, Action: Forward) This is a simple pass-through connection between input and output ports.	Mempool create Link create Pipeline create Pipeline port in/out Pipeline table Pipeline port in table Pipeline enable Pipeline table rule add
2.	Flow Classification (Table: Exact match, Action: Forward)	Mempool create Link create Pipeline create Pipeline port in/out Pipeline table Pipeline port in table Pipeline enable Pipeline table rule add default Pipeline table rule add
3.	Firewall (Table: ACL, Action: Forward/Drop)	Mempool create Link create Pipeline create Pipeline port in/out Pipeline table Pipeline port in table Pipeline enable Pipeline table rule add default Pipeline table rule add
4.	IP Routing (Table: LPM, Action: Forward)	Mempool create Link create Pipeline create Pipeline port in/out Pipeline table Pipeline port in table Pipeline enable Pipeline table rule add default Pipeline table rule add

5.6 CONCLUSION

The legacy network appliance and systems are often designed for a specific workload, which may use the purpose-built silicon which is not able to be repurposed as service-oriented software-defined infrastructure. System application and service can be more easily implemented and upgraded on the off-the-shelf server with software approach. In order to do it, the packet forwarding model, pipeline, and algorithms are the important choices to build the performant networking workload. DPDK supports the RTC, pipeline, and hybrid models which are fundamental design paradigms for a system architecture. With all these software ingredients, the network system can be built more easily and quickly by software networking developers.

6 PCIe/NIC

Cunming Liang, Hunt David, and Heqing Zhu

CONTENTS

Now, the focus moves from CPU to I/O device. This chapter discusses the PCIe bus that connects CPU and I/O, including the packet movement procedure between CPU and NIC, along with a quantitative analysis of the PCIe bandwidth usage. This chapter also explores the memory buffer and its performance optimization practices.

6.1 PCIe

6.1.1 PROTOCOL STACK

PCIe (PCI-e) stands for Peripheral Component Interconnect Express; it is a high-speed serial communication and interconnect standard interface. Today, it is the de-facto interface to connect flexible I/O devices to the open server platform; as a matter of fact, the system bus is the bus interface to CPU, which is not open interface, which leaves the PCIe as the platform standard bus interface.

The PCIe standard specification is maintained by the PCI Special Interest Group (PCI-SIG). PCIe replaced the legacy bus architectures such as PCI, PCI-X, and AGP.

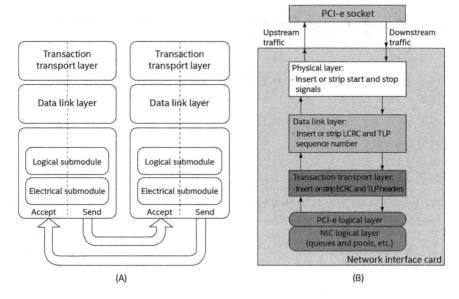

FIGURE 6.1   PCIe protocol stack, NIC perspective.

The server platform can support many PCIe interfaces for flexible I/O use and system expansion. The PCIe protocol stack can be broken into three layers, as shown in Figure 6.1:

- Transaction transport layer;
- Data link layer;
- Physical layer (logical and electrical submodules).

A NIC is the network interface card on a server platform, and it is mainly connected to a switch/router with a wired networking interface. Today, most wireline NICs are PCIe-based cards. In client platforms (desktop, mobile), some low-cost NICs are built with a USB interface and are not yet supported by DPDK. Wireless NICs are also not yet supported by DPDK, but this functionality is under development.

For a specific NIC, if PCIe is used as the interface (physical socket) to connect to the server platform, the inbound traffic going from NIC (the physical layer) to the PCIe root complex (server platform) is called upstream, while the outbound traffic going from the PCIe root complex to the NIC (the physical layer) is known as downstream. When incoming packets arrive on the NIC, the packets go through the PCIe bus transaction transport layer, and then move into system memory (or sometimes direct to CPU cache). The incoming packets (single or multiple) are inserted into "TLP (transaction-layer packet) data" using format shown in Figure 6.2.

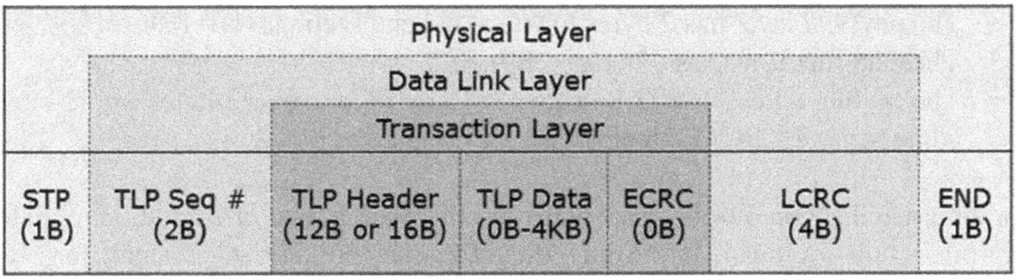

Physical Layer						
	Data Link Layer					
		Transaction Layer				
STP (1B)	TLP Seq # (2B)	TLP Header (12B or 16B)	TLP Data (0B-4KB)	ECRC (0B)	LCRC (4B)	END (1B)

FIGURE 6.2 PCIe layers.

6.1.2 Transaction Overhead

The TLP format is used for the data transport over PCIe. It has a header, data, and checksum sections. The TLP header section defines a set of transactional packet types, which are listed in Table 6.1.

The packet data is carried as the TLP payload in the transport layer. The Ethernet packets that the NIC receives the wire are transferred as the payload of the PCIe transport layer. Generally, the NIC implements its own DMA (direct memory access) controller to access the memory via the PCIe bus. When the CPU receives/sends a packet from/to the NIC, the following PCIe transaction types are used: Memory Read/Write (MRd/MWr) and Completion with Data (CpID). As the incoming packet is used as part of the "TLP data", it raises the question of how much the extra "bytes" overhead is added on a PCIe-based transaction? Is it significant?

TABLE 6.1
TLP Types

TLP Types	Abbreviated Name
Memory read request	MRd
Memory read locked access	MRdLk
Memory write request	MWr
I/O read	IORd
I/O write	IOWr
Configuration read (Type 0 and Type 1)	CfgRd0 CfgRd1
Configuration write (Type 0 and Type 1)	CfgWd0 CfgWd1
Message request without data	Msg
Message request with data	MsgD
Completion without data	Cpl
Completion with data	CpID
Completion without data associated with locked memory read requests	CplLk
Completion with data associated with locked memory read requests	CplDLk

- The physical layer has 2 bytes for start and end sections (1B+1B).
- The data link layer has 6 bytes. (2B+4B).
- The 64-bit addressable TLP header is 16 bytes long (or 12 bytes for 32-bit addressing). ECRC (Endpoint CRC) is optional as 0B.

For a normal TLP on 64-bit server platform, there are a total of 24 additional bytes for transaction overhead. If the incoming Ethernet packet size is small (e.g., 64 bytes), the overhead is indeed significant.

6.1.3 INTERFACE METRICS

PCIe technology has gone through multiple iterations in the past 15 years. As of today, a typical PCIe device is often specified as follows:

- Supports PCIe Gen1/2/3/4/5. Each version increases performance significantly.
- Number of PCI lanes per slot, often specified as x1/x2/x4/x8/x16. More lanes mean more performance.

For example, Intel® XL710 NIC supports PCIe Gen3 x8. Intel® 82599 NIC supports PCIe Gen2 x 8 and PCIe Gen3 x4. PCIe Gen4/Gen5 is the latest technology available for extremely high-performance use cases. Gen2 doubles the single-lane transfer rate of Gen1. With the improvement in coding efficiency, the actual effective single-lane data rate is nearly twice as fast as that of Gen2. Gen1 and Gen2 use an 8 b/10 b encoding scheme, which means only 80% of the transfers across the bus is data and 20% is overhead. Gen3/4 are improved with 128b/130b encoding scheme, and the overhead is less than 1% (Table 6.2).

The bit rate of data across the PCI bus is measured in GT/s (giga-transfers per second, where a transfer is one bit across the PCI bus). For 8 b/10 b encoding, 8 bits (1 byte) of data needs 10 bits transferred across the bus. GT/s can be estimated into throughput together with encode mechanism; Gb/s is the effective data rate in the physical layer, indicated by the last column. It can multiply with number of lanes to increase I/O capability. If a NIC supports PCIe Gen2 x8, which means 5 GT/s, a single-lane throughput is 4 Gb/s; with 8 lanes support (x8), the maximum throughput

TABLE 6.2
PCIe Throughput in Generations

Version	Year	Encode Mechanism	Transfer Rate	Throughput Per Lane (Per Direction)
Gen1	2003	8b/10b	2.5 GT/s	250 MByte/s, 2 Gbit/s
Gen2	2007	8b/10b	5.0 GT/s	500 MByte/s, 4 Gbit/s
Gen3	2010	128b/130b	8.0 GT/s	984.6 MByte/s, 7.88 Gbit/s
Gen4	2017	128b/130b	16.0 GT/s	1969 MByte/s, 15.75 Gbit/s

can be estimated as 32 Gb/s per direction. This is the throughput from the PCIe physical layer, and this is not the data rate for networking port speed (data rate) yet. As shown earlier, each PCIe transaction is done with the PCIe protocol, which is a packet-based bus transaction. There are 24 bytes overhead for each TLP data transaction. If the TLP data is 64 bytes long, the effective packet transfer ratio is estimated as 73%.

```
(1-64 / (24 + 64)) ~ = 73%.
```

How do we translate the physical throughput of a slot into I/O data rate, and can PCIe Gen2x8 support 2x10 Gbps? It looks like it will according to the following equation:

```
32 Gb/s * 73% ~= 23 Gbps
```

Unfortunately, this is not a complete story as there is more overhead to take into consideration in the PCIe transaction. On a Linux box, it is easy to list out PCIe devices: "lspci" is such a command to do it. Here is an example to read the device capability on a PCIe Gen2 x8 NIC.

```
Capabilities: [a0] Express (v2) Endpoint, MSI 00
              DevCap: MaxPayload 512 bytes, PhantFunc 0,
              Latency L0s <512ns, L1 <64us
                  ExtTag- AttnBtn- AttnInd- PwrInd- RBE+
                  FLReset+
              DevCtl: Report errors: Correctable+ Non-Fatal+
              Fatal+ Unsupported+
                  RlxdOrd+ ExtTag- PhantFunc- AuxPwr-
                  NoSnoop+ FLReset-
                  MaxPayload 256 bytes, MaxReadReq 512
                  bytes
              DevSta: CorrErr+ UncorrErr- FatalErr-
              UnsuppReq+ AuxPwr- TransPend-
              LnkCap: Port #2, Speed 5GT/s, Width x8, ASPM
              L0s, Exit Latency L0s <1us, L1 <8us
              ClockPM- Surprise- LLActRep- BwNot- ASPMOptComp-
              LnkCtl: ASPM Disabled; RCB 64 bytes
              Disabled- CommClk+
                ExtSynch- ClockPM- AutWidDis- BWInt- AutBWInt-
              LnkSta: Speed 5GT/s, Width x8, TrErr- Train-
              SlotClk+ DLActive- BWMgmt- ABWMgmt-
              DevCap2: Completion Timeout: Range ABCD,
              TimeoutDis+, LTR-, OBFF Not Supported
              DevCtl2: Completion Timeout: 50us to 50ms,
              TimeoutDis-, LTR-, OBFF Disabled
         LnkCtl2: Target Link Speed: 5GT/s, EnterCompliance- SpeedDis-
                  Transmit Margin: Normal Operating
                  Range,EnterModifiedCompliance-ComplianceSOS-
                  Compliance De-emphasis: -6dB
```

```
          LnkSta2: Current De-emphasis Level: -6dB,
EqualizationComplete-, EqualizationPhase1-
                      EqualizationPhase2-,
EqualizationPhase3-, LinkEqualizationRequest-
```

6.2 TLP EXAMPLE

In addition to the TLP format, the PCIe implementation may also add more overhead. For example, some NICs may require each transaction (TLP) to start from Lane 0, or from an even number of clock cycles. These limitations, while not mandated by the PCIe specification, are subject to the specific vendor implementation, which may further reduce the effective throughput. Here is a small experiment on a simple write of 64 bytes of data, which is done on a Gen2 x8 interface. A PCIe protocol analyzer captures the actual PCIe transaction. Let's analyze the effective bandwidth in the upstream direction. The impact of the specific hardware implementation is also considered. So, a write of 64 bytes requires 12 clock cycles, as shown in Figure 6.3.

At the physical rate of 4 GB/s, 8 bytes can be transferred per clock cycle in 2 ns, giving a total of 24 ns for 12 clock cycles. Only 64-byte effective data can be transferred in 24 ns, so the effective transfer rate is around 2.66 GB/s (64B/24 ns). Compared to the physical bandwidth of 4 GB/s, this is only about 66.6% (64B/96B) of the effective data write bandwidth. This is lower than 73%, and it is caused by the additional implementation overhead (start from the even number of clock cycles).

This is an example of a PCIe memory write. Typically, the packet write of NIC is done by DMA. Prior to the actual data (memory) write, there are a series of control actions before the TLP data transfer (packet write). These actions will further affect the PCIe interface utilization, and each action consumes the bandwidth. Let's take a closer look at PCIe TLP—the application layer packet transmission is managed by the NIC's DMA controller.

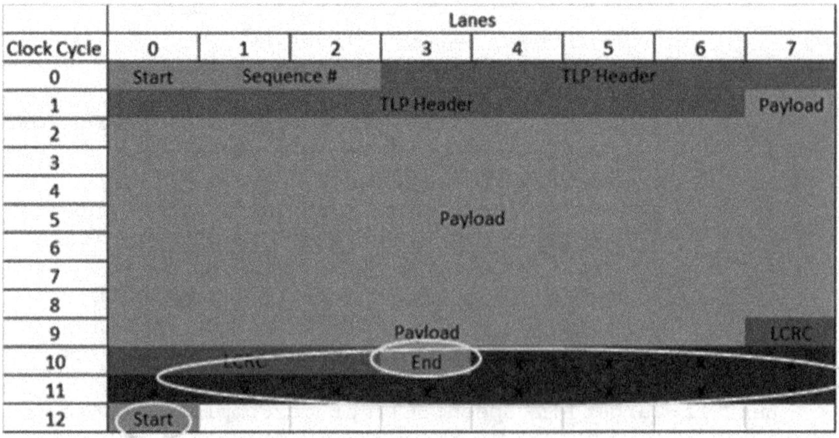

Clock Cycle	Lanes							
	0	1	2	3	4	5	6	7
0	Start	Sequence #			TLP Header			
1			TLP Header					Payload
2								
3								
4								
5			Payload					
6								
7								
8								
9			Payload					LCRC
10		LCRC		End				
11								
12	Start							

FIGURE 6.3   TLP format: 64-byte memory write.

6.3   NIC DESCRIPTOR

DMA is a high-speed data transfer mechanism which allows data reads and writes between the external device and the system memory. Data will not go via the CPU nor be touched by the CPU. The whole data transfer is managed by the DMA controller on the device. This allows the CPU to work on other tasks during the data transfer process. Some NICs have built in with DMA controllers, allowing them to follow this usage model.

The NIC interacts with the CPU (and its software) via a ring-based queue, which consists of many packet buffer descriptors, with the ring being a circular buffer in system memory. Usually, each packet has its own memory buffer and an associated packet buffer descriptor. That means each descriptor refers to the buffer address of the packet memory. The queue has a few key control registers such as **Base, Size, Head,** and **Tail** pointers. The CPU can allocate a segment of physically consecutive memory, and the start address is assigned into the **Base** register of the queue. The **Size** register is configured with the number of all buffer descriptors. The **Head** register is often read-only for the software, while the NIC can write this **Head** register, which refers to the descriptor that is currently in use. The **Tail** register is owned by the software driver. At the system initialization phase, the NIC driver will configure all the device settings, and the NIC DMA controller is notified that the queue is configured with a list of packet buffer descriptors (Figure 6.4).

Figure 6.4 illustrates the descriptors on the queue, with the example being based on Intel® 82599 NIC. The NIC can access all the descriptors between **Head** and **Tail**.

The queue is empty if the **Head** equals the **Tail,** which means no packets have been received. The queue is full if **Head** equals Next (**Tail**). Every entry in the queue is a packet buffer descriptor. The descriptor format and its size are NIC specific. For example, a descriptor size on Intel® 82599 is 16 bytes. The NIC descriptor details are usually available in the NIC data sheet. The NIC driver will follow the hardware definition. Following DPDK memory optimization principles, the buffer queue size will usually be a multiple of the largest cache line. The start address of the queue should also be cache line-aligned. The NIC can work with the different driver models, interrupt mode, poll mode, or both.

Each packet buffer descriptor has a bit to indicate "Descriptor Done", which tells if the packet has been received or sent successfully, and the software driver must check and use this bit. The NIC's DMA controller writes back the "status" bit to its descriptor, and this is a signal that NIC has completed all operations on the specific packet, and that it is time for software driver to use it. Regardless of the packets RX and TX, the NIC driver is responsible for the following basic steps:

1. Fill the packet memory address into the descriptor so that the NIC can use it to DMA the packet arrives on wire.
2. Move the **Tail** register; NIC will read and know how many descriptors are available.
3. Check the "DD" bit of the descriptor. For packet RX is done, refilling the packet memory buffer is required. For packet TX is done, releasing the memory buffer is done.

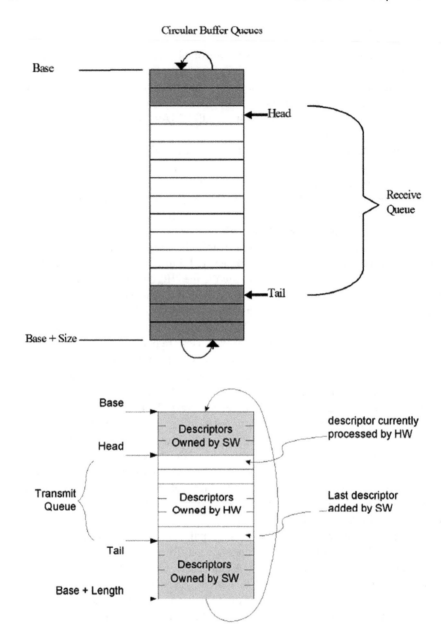

FIGURE 6.4   Intel® 82599's descriptor.

In addition to the basic steps, the NIC driver is also responsible for the following steps:

- Configure the buffer descriptor (rte_mbuf).
- Process the buffer descriptor.
- Handle the scatter-gather, RSS (receive side scaling) and NIC offload, etc.

DPDK is different from the traditional NIC drivers, in that it is designed for optimal processing of packet RX/TX. It is important to understand the basic steps and then proceed to the optimal combinations. It is important to note that the NIC driver is not computationally intensive, but I/O intensive.

From the CPU perspective, the NIC driver consists of a small amount of computation, a large amount of memory access, and some MMIO (memory-mapped IO)-related actions. The software optimizations aim to complete packet I/O using the minimal execution time, allowing it to handle more packets in a given time. The CPU optimizations are only one aspect of the optimizations. In addition to the CPU software optimizations, I/O optimizations also play an important role. They determine how many packets (data) can be processed by the CPU. As described earlier, a NIC is plugged in via PCIe interface. The PCIe protocol stack and transaction efficiency are important; TLP and its application data encapsulation add the additional overhead; and the implementation and usage efficiency determine the actual effective use of PCIe bus bandwidth.

The platform memory bandwidth is usually significantly higher than any individual PCIe interface bandwidth. With that in mind, Direct Data I/O (DDIO) on Intel® processors is another technical improvement from the memory perspective, as it can reduce the memory access and it can place the packet into cache directly. For packet processing workloads, memory bandwidth is not a big challenge; it is mainly the PCIe interface that is the bottleneck.

For storage-intensive workloads, there are other concerns, such as the total memory demand and the effective utilization of the memory bandwidth. This is not discussed in this chapter.

6.4 RX/TX PROCEDURE

Once the DMA controller finishes the packet I/O using the packet buffer descriptor and the queue mechanism, the software driver handles the remaining tasks running on the CPU. All the packet memory buffers, descriptors, and the queues are located in the system memory. NIC control is done by register access.

From a CPU perspective, there are packet-related memory accesses (including the cache access with Intel® DDIO) and MMIO accesses. The access to the device register has to go via MMIO, which goes through the PCIe interface, and as such, it is quite slow and should be avoided where possible. One optimization idea is to improve the overall PCIe utilization; actions should be minimized during packet I/O. PCIe bus access latency is also high. Overall, MMIO optimization is important, and it will be discussed later.

From a NIC perspective, the DMA controller on the device is responsible for accessing the system memory and NIC memory. The system memory access has to go through the PCIe transaction, which is the effective packet movement; this large data movement consumes the most of PCIe bandwidth. It is therefore essential to understand these detailed steps.

6.4.1 HIGH-LEVEL VIEW

To achieve high performance on packet RX/TX, it is essential to understand all the fundamental steps in the platform hardware. Understanding the basics helps to make subsequent software-centric performance-tuning efforts. If the NIC configuration via device registers is left aside for a moment, the key focus must stay on the packet sending and receiving, and the packet data movement paths among the CPU, memory, and NIC. The DDIO technique of Intel® processors allows the NIC's DMA engine to move packets into the LLC (last level cache) directly. In the following example, packet processing ends at the LLC (Figures 6.5 and 6.6).

Figures 6.5 and 6.6 describe the detailed packet RX/TX procedure. Each step is indicated with a sequence number, which may not indicate all execution steps in order. For example, the packet memory refill on the RX side and the memory release on the TX side are not included here.

6.4.1.1 RX

1. The CPU fills the packet buffer address into the RX descriptor.
2. The NIC reads the RX descriptor (rxd) to obtain the memory address.
3. The NIC writes the packet to the specified memory address (DMA).
4. The NIC updates the RX descriptor to notify the RX completion.
5. The CPU reads the RX descriptor and completes the packet RX

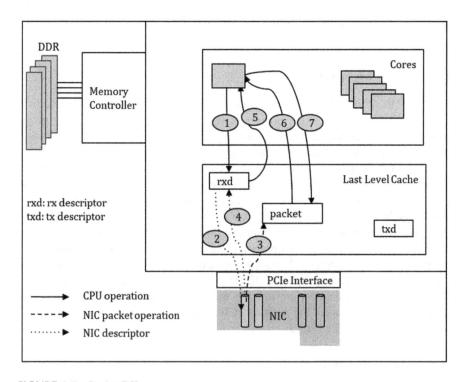

FIGURE 6.5   Packet RX.

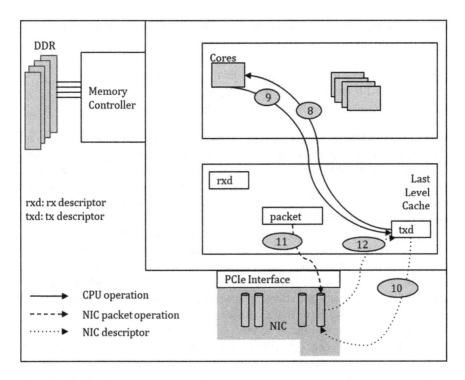

FIGURE 6.6 Packet TX.

Steps 1 and 5 are the memory access, CPU reads and writes the data from memory to LLC (last level cache);

Step 2 is the PCIe downstream operation;

Steps 3 and 4 are the PCIe upstream operations.

6. The CPU processes the packet and decides the next step (processing).

7. The CPU may change the packet and send it.

Steps 6 and 7 are beyond the NIC processing. It is still packet processing, mostly memory access that is relevant without using PCIe.

6.4.1.2 TX

8. The CPU reads the TX descriptor to check if any packet is sent successfully.

9. The CPU writes the packet buffer address to the next TX descriptor.

10. The NIC reads the TX descriptor to get the packet memory.

11. The NIC reads the packet data from the address.

12. The NIC updates the TX descriptor after packet TX is completed.

Steps 8 and 9 are the memory access, and CPU reads and writes to LLC;

Steps 10 and 11 are the PCIe downstream operations;

Step 12 is the PCIe upstream operation.

In the above steps, the memory access and PCIe-related access are purposely separated and highlighted.

- Step {2,10,11} are the PCIe downstream operations.
- Step {3,4,12} are the PCIe upstream operations.
- The remaining steps are related to memory access.

If NIC control (MMIO access to device registers) is taken into account, there are more PCIe downstream operations. One thing to note here, the request and the completion are often in pairs. During the PCIe transaction, the request in the downstream direction can go together with the completion in the upstream direction. Overall, there is more pressure on the PCIe upstream direction, as shown by PCIe protocol analysis and transaction tracing.

Analyzing the maximum theoretical bandwidth of the interface is important for performance tuning and optimization. This provides a quantitative guidance on optimization headroom. From the early analysis, the physical bandwidth of PCIe interface is indeed high. Take PCIe Gen2 x8 as an example: The throughput is up to 4 GB/s (32 Gbps). In reality, the effective data throughput is much less. Why? Because the packet moves between the main memory and the NIC device through DMA and PCIe bus transactions, and the PCIe bus is consumed by in/out packet data movement and NIC register access (MMIO). Also, the reads/writes {2,4,10,12} of the RX/TX descriptors described above need to be taken into consideration.

We need to go through all steps to understand the transaction overhead of every step, so that we can calculate the theoretical bus utilization bandwidth. An example will be presented below.

6.4.2 OPTIMIZATION IDEAS

Up to now, NIC RX/TX and CPU/SW drivers have been introduced. The steps are a collaborative procedure that allows hardware and software components to work together. This still leaves room for the software developer to create a different software driver, so let's explore the possible optimization approaches with the aim of achieving the highest PCIe bus utilization possible.

1. Reduce the MMIO access:

    NIC configuration and control are done by accessing the device registers; they use MMIO access, and frequent MMIO access slows down performance. MMIO access involves register access of the TLP data payload, and in this case, the packet overhead is much more than the actual data payload on MMIO-based register access.

    When a packet is received, the **Tail** register is updated after a new packet buffer is allocated and refilled to the RX descriptor. If the packet buffer allocation and descriptor refills are batched together, we can reduce the total time spent updating the **Tail**. For example, if we only update the tail once every four descriptors, the **Tail** access is effectively reduced from 4 to 1. DPDK takes advantage of this by using a buffer threshold at RX side, where the threshold is related to how many packet buffers are available for packet RX. As long as the threshold is not hit, which means there are enough packet buffers available for NIC use, the Tail update can cover many descriptors, reducing the MMIO access frequency, but not affecting the actual packet RX functionality. This method is only applicable to RX side; it is not usable for TX side. The tail update is used to notify NIC to send packets, and any delay in updating the Tail pointer will slow down the TX request to NIC. However, if there are many packets to be sent simultaneously, then the packet buffer allocation and TX descriptor updating can happen in a bulk fashion, which again reduces the total number of MMIO accesses.

2. Improve the PCIe data transfer efficiency:

Packet descriptors can be accessed by both the CPU and the NIC. Every read or write of the descriptor is a PCIe transaction. Assuming each descriptor is 16 bytes long, then the payload will be smaller than PCIe/TLP overhead, resulting in a lower utilization on PCIe/TLP data transfer bandwidth.

If we combine multiple descriptor accesses, we can merge four descriptor accesses into one PCIe transaction, which will result in a total payload of 16 bytes $\times$ 4 = 64 bytes, which happens to be the same as the cache line size. So, this larger TLP payload can make much more effective use of PCIe bandwidth.

On the TX side, the NIC will write back the status bit to descriptor after the successful packet transmit. Usually, this happens for every packet. The NIC can be configured in a way so that the write-back is coalesced for multiple packets. For example, on Intel® 82599 NICs, RS (Report Status) bit is part of NIC descriptor, and it can be set to enable TX write-back batching. This approach allows the NIC to confirm the successful packet transmit in a bulk fashion, where it can write back once after multiple packets are sent out.

3. Avoid partial writes of the cache line whenever possible

When a DMA engine is writing the packet data to the memory buffer, partial writes can happen in the following conditions:
- The buffer address is not cache line-aligned.
- The written length is not a multiple of the cache line size.

Partial writes of the cache line will lead to a merge of memory read–modify–write operations and cause additional reads, thus reducing the overall performance. So, it is necessary to make sure all packet buffer addresses are cache line-aligned, which is another optimization implemented in DPDK.

A simple example would be where a 64-byte packet arrives; if NIC is configured with CRC strip, then the total packet size is reduced to 60 bytes, and the RX/TX performance result is worse than 64 bytes. This can be easily observed in the real test environment.

The aforementioned optimizations may depend on NIC support, and the key principles should be easy to understand. The PCIe bus transactions are part of a packet-based protocol stack, and its effective utilization is important. The NIC design goes to great lengths to improve the transaction efficiency at the hardware support, but the software driver needs to take advantage of its hardware tuning capabilities.

6.5 THROUGHPUT ANALYSIS

So far, the packet RX/TX paths have been discussed, and the CPU and NIC have their own role and responsibilities via PCIe transactions, so now let's look at the theoretical throughput of a PCIe-based NIC. The Intel® 82599 NIC's I/O interface, e.g., supports 2x10GE ports. The host interface is based on PCIe Gen2x8, which provides 4 GB/s in a single direction. In most common use cases, the upstream transaction will have more pressure than the downstream transaction. This analysis will focus on what happens in the upstream direction when dealing with small packets. 64-byte packets will impose the highest pressure to the PCIe bus. PCIe upstream direction consists of a list of operations:

- Write access listed in steps {3,4,12}.
- Read access listed in steps {2,10,11}.

The data access size in each step is determined by the packet size, so all the aforementioned steps can be combined to calculate the total number of data (in bytes) going across the PCIe bus in order to forward a single packet.

- The NIC writes the RX packet data to the system memory (Step 3)
    Intel® 82599 requires TLP transaction that must align with an even number of cycles, so indeed 64-byte packet payload does occupy 96 bytes (12 cycles totally) for PCIe transaction.
- The NIC writes back RX descriptor (Step 4)
    When we want to write back the whole cache line in 82599 (64 bytes), this will occupy 96 bytes in the PCIe transaction. With the batched descriptor refilling optimization, a cache line can hold up to four descriptors in total. A single packet only needs 24 bytes, 96 bytes / 4 = 24 bytes.
- The NIC writes back TX descriptor after packet TX (Step 12)
    The RS bit is used on the TX side; write-back can be tuned to happen once every 32 packets. The overhead on a single packet is close to negligible.
- The NIC reads the RX descriptor (Step 2)
    For a read request, the TLP data section is null, and only 24 bytes of PCIe overhead is required. Intel® 82599 supports the TLP to align with an even number of clock cycles; it will occupy 32 bytes. Likewise, four descriptors can be merged into one read access. On average, one single packet takes 8-byte overhead, i.e., 32 bytes/4 = 8 bytes.
- The NIC reads TX descriptor (Step 10)
    Similar as above, i.e., 8 bytes.
- The NIC reads the packet data for TX (Step 11)
    Each packet has a read request, i.e., 32 bytes.

In summary, moving a single 64-byte-long packet requires 168 bytes = (96 + 24 + 8 + 8 + 32) of PCIe data traffic. The maximum upstream forwarding rate of a 64-byte packet is then estimated as 23.8 Mpps on an Intel® 82599 NIC.

$$4000\,\text{MBytes} / 168\,\text{Byte} = 23.8\,\text{Mpps}$$

This is less than 2 × 10 Gbe port, and 10 Gbe line rate is about 14.88 Mpps. The expected I/O forwarding rate is indeed 29.76 Mpps.

$$14.88\,\text{Mpps} \times 2 = 29.76\,\text{Mpps}$$

In the 64-byte packet scenario, the NIC cannot reach the line rate (2 × 10 Gbps) as it is limited by the PCIe interface overhead. For the larger packet scenario, this limit no longer exists. PCIe transaction has higher efficiency for larger packets as the overhead is proportionally smaller. Since Internet traffic is a mix of many packets in different sizes, it's worth using a PCIe protocol analyzer to inspect all PCIe-relevant transactions when tuning. Different NICs can have subtle differences in implementation, and the math model might be slightly different to that described above.

6.6   PACKET MEMORY

The optimization ideas in the previous sections depend on an effective and efficient implementation. DPDK designed the packet-based memory buffer descriptor using the software optimization principles, and it is based on FreeBSD memory buffer design.

6.6.1   MBUF

Every packet needs a place in memory, commonly known as packet memory (or packet buffer). DPDK refers to this memory as an mbuf. The mbuf (struct rte _ mbuf) is designed as a packet-based data structure. The mbuf layout is described below, and contains the actual packet data and its buffer descriptor (packet metadata).

The mbuf header is known as the packet descriptor (or buffer descriptor), and its size occupies two cache lines. The software driver uses this to interact with NIC hardware for every packet in and out. The same data structure can be used for hardware accelerators such as crypto or FPGA, but it is generally designed for packet-processing tasks. The most frequently used data fields are in the first cache line (most commonly accessed), and the extended packet-processing fields are located in the second cache line. The mbuf header is metadata, in that it describes the packet, and software driver may use the NIC's hardware-assisted capability to gather the packet metadata. The packet data is placed later in the mbuf. All mbufs are cache line-aligned, but their size is configurable. Given that most Ethernet frame sizes are less than 1500 bytes, it is common to see the default mbuf packet size is set as 2KB (Figure 6.7).

As part of the design pattern, packet descriptors and packet data can be placed in the same memory region. Alternatively, they may use different memory regions,

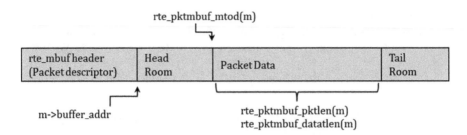

FIGURE 6.7   rte_mbuf's layout.

different allocation methods, and an alternative management interface. At initialization, the DPDK application usually chooses one particular memory scheme for the packet descriptors and data, simplifying implementation and management.

The head room and tail room sections of the mbuf structure are available for packet modification, and the room can be reserved to save the need of memory copy. Most packet modifications happen in the packet head (such as adding/removing new protocol headers) and/or packet tail (such as adding/validating the packet integrity), the packet copy can be kept in the same place, the head/tail change can be done using the reserved rooms, and the pointer can be moved accordingly. The length of the headroom may be defined by RTE_PKTMBUF_HEADROOM. Packet descriptor (metadata) contains the packet header information such as VLAN label, RSS hash values, and arrived port number.

If the received packet is a jumbo frame (its size is larger than 9KB), it may occupy multiple mbuf elements. DPDK supports this buffer chaining, where the packet is stored in multiple chained mbuf elements. Support for Ethernet jumbo frames is important for many network systems; though, in reality, regardless of whether jumbo frame is supported effectively, it is up to the deployed network nodes. It is common to see the packet segmentation in the transit node, and the packet reassembly then needs to happen in the network endpoint. For jumbo frames, packet metadata is used in the first mbuf element, but not required in the subsequent chained mbuf elements (Figure 6.8).

The starting address of the headroom is pointed by the **buff_addr** of rte_mbuf. In lib/librte_port/rte_port.h, a macro is available for obtaining the pointer and special offset data from buff_addr. Please refer to macros such as **RTE_MBUF_METADATA_UINT8_PTR** and **RTE_MBUF_METADATA_UINT8** in rte_port.h source code.

The **rte_pktmbuf_mtod (mbuf)** function can get the packet data starting position. For a single packet in a single mbuf, the data length is the same as the packet length, and it can be obtained from **rte_pktmbuf_pktlen (mbuf)** or **rte_pktmbuf_datalen (mbuf)**. The jumbo frame length must be obtained by **rte_pktmbuf_datalen (mbuf)**, along with the sum of the data lengths of the chained mbuf elements.

Memory buffers can be used for packet or control purposes, and the request is sent to DPDK mempool. Mbuf allocation and initialization is responsible for memory allocation and the initial setup, such as the mbuf type, memory pool, and cache/core-related address.

Packet (m) = mbuf(m1) + mbuf(m2) + mbuf(m3)

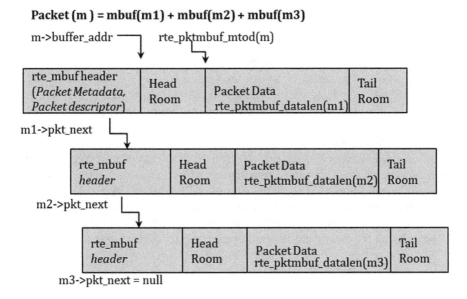

FIGURE 6.8 Jumbo frames in chained mbuf.

- *rte_pktmbuf_alloc()*: This is used to allocate memory for the new packet.
- *rte_pktmbuf_init()*: This is used to initialize the mbuf.
- *rte_ctrlmbuf_alloc ()*: This is intended to create the control-related memory buffer. Initializing the mbuf can be done by rte_ctrlmbuf_init().

Releasing the mbuf is to return the memory back to the mempool; there are core and cache-related contents, which will not be initialized before a new allocation. The general mbuf-related interface are listed below:

- *rte_pktmbuf_datalen()*: Get frame data length.
- *rte_pktmbuf_mtod()*: Get the packet/data pointer.
- *rte_pktmbuf_prepend()*: Insert data before the packet.
- *rte_pktmbuf_append()*: Append the data after the packet.
- *rte_pktmbuf_attach()*: Attach the memory buffer, which connects the packet data and its descriptors that are placed in different memory area, known as the use of the indirect buffer. Accessing an indirect buffer is not efficient as accessing a direct buffer (the typical rte_mbuf use case).
- *rte_pktmbuf_detach()*: Detach the memory buffer.
- *rte_pktmbuf_clone()*: Clone the mbuf.

6.6.2 MEMPOOL

All packets use mbufs, which are managed through the DPDK mempool interface. The mempool design is based on a ring concept, which manages the used/free mbuf objects for packet RX/TX.

When a packet arrives on a NIC, the DPDK PMD requests the mbuf object and informs NIC of its memory location. Indeed, the mbuf object is just taken from the mempool, and it is a ring of mbuf. This is not a system memory allocation from OS. At DPDK initialization phase, the memory pool has been reserved and created, and mbuf is part of the mempool. After the mbuf object is taken by the PMD, the memory object will be updated with the packet-specific metadata such as Ethernet and IP protocol information. Once the packet is sent out, the used mbuf will be returned back for future use. This does not require the dynamic memory management from OS, and DPDK is responsible for it.

When multiple cores are needed to drive the NIC interface to its full potential, the mempool and its rings are shared data among these cores as each core will request and release mbuf objects. The shared data use will result in lower efficiency, with each memory access requiring a "compare-and-set" operation. This is to ensure that the same data will not be modified by multiple cores. DPDK resolves this problem by introducing the local cache concept, where each core has its own local cache of mbufs. This reduces the overall frequent access to the mempool ring buffer, but comes at a cost of increased memory usage.

6.7   CONCLUSION

In this chapter, the PCIe interface, DMA, descriptor, and packet movement techniques and their optimizations were discussed. The packet descriptors and their related memory buffers in the context of NIC RX/TX were also discussed. Further details of DPDK mbufs and mempools can be found in the online documentation. In the coming chapters, more NIC-related content will be introduced.

7 PMD

Helin Zhang and Heqing Zhu
Intel®

CONTENTS

PMD is one of the most valuable technology contributions from DPDK. It takes the advantage of PCIe transaction and memory buffer optimizations shared in Chapter 6. PMD is a new way of implementing a NIC driver. This chapter also discusses the optimization examples such as NIC configuration and performance-tuning tips.

7.1 DPDK POLL MODE

Prior to DPDK, the traditional NIC drivers are running in the kernel space and usually use the interrupt mode for packet RX/TX. DPDK earned its reputation due to the user space poll mode implementation; this effectively drives the high-speed NIC. In the busy poll mode, the dedicated core(s) can be assigned to focus on packet RX/TX. DPDK PMD works in Linux user space, and it is closer to DPDK-based application, which mostly runs in the user space.

If the application runs in the kernel space, how about to move PMD to the kernel space? This idea has been implemented by some developers; the migration efforts and results have been presented in one of DPDK Summits, though the code is not shared in public domain. As of 2019, some Linux kernel-based NIC drivers have been implemented with XDP (eXpress Data Path), which can do the busy poll or interrupt model at the kernel space. DPDK also supports the interrupt mode, and the hybrid use of poll and interrupt modes for packet RX/TX.

7.1.1 INTERRUPT MODE

When a packet enters the NIC, it goes into one of RX queues. The NIC will generate a hardware (MSIX/MSI/INTX) interrupt signal, the interrupt is routed to a CPU, and then an interrupt service handler is responsible for receiving the packet. The interrupt service has the cost of context switch, and the overhead exists for every interrupt service. When a CPU runs at a much higher rate than I/O device, such overhead can be affordable to some extent, and it can even be ignored. Interrupt is asynchronous; CPU can work on other tasks when there are not many packets to enter the system. Depending on the actual arrived packets in a short period, interrupt can be very effective.

But if a high-speed NIC is connected, large burst incoming packets lead to many interrupts, the NIC and software driver can take care of the burst packets with less interrupt and service, and one interrupt service routine can handle many packets, which reduces the system overhead. Still, the large amount of data gets in and out of the computer system when NIC is doing the packet RX/TX in the line speed; the cumulative overhead increases to a significant level, which becomes a problem that needs to be addressed.

When a packet needs to be sent out, the NIC driver will prepare the packet memory and set up the NIC descriptor on the TX queue using the buffer descriptor. When the packet TX is completed, the NIC also generates a hardware interrupt, and CPU will use the interrupt handler to complete the remaining tasks after packet TX, like memory buffer release. The TX interrupt has the similar context switching overhead as packet RX. Interrupt mode is effective if there is no high demand on packet RX/TX.

7.1.2 POLL MODE

DPDK's poll mode refers to use the dedicated core resource for packet processing, and the core is running in the busy poll mode, without using the interrupt service mechanism. The busy poll mode is more effective in a high-throughput scenario. Without using the interrupt mode, it saves the context-switching cost. Through the math model and analysis, it makes sense to assign the dedicated cores to handle the high-speed NIC in order to sustain the packet processing at the line rate.

In this mode, the interrupt can be disabled at NIC port initialization phase. The entire process of packet RX/TX is described in detail in Chapter 6. Any packet entering the NIC will go through the hardware check, which consists of integrity check, header parsing, and packet filtering before it finally enters an RX queue of a port. In general, every RX queue is paired with a software queue configured with mbuf elements. PMD is responsible for NIC port/queue initialization and setup, which includes filling the mbuf elements to HW (Hardware) descriptor and informing NIC to place the incoming packet to the specified memory address; packet metadata is updated into the buffer descriptor. PMD is responsible for allocating more buffers to accommodate the large size packet.

If NIC completes the packet RX, it will update the "DD" bit of the HW descriptor software will keep checking this bit. If set, the software will determine a new packet that is now ready in the corresponding packet buffer, and PMD will read the NIC HW descriptor and update the packet-relevant information from HW into the buffer descriptor, which is part of mbuf. A new packet buffer will be allocated to this NIC HW descriptor in order to receive the next packet. HW descriptor is specific to each NIC, and the buffer descriptor is general interface in DPDK, which works for all NICs, even if they are built by the different suppliers.

It is a similar procedure as the TX side; every TX queue is paired with a software queue; PMD is responsible for the queue setup and management. System designer will assign cores to the specified TX port/queue. PMD will read the mbuf and use the packet buffer descriptor to set up the NIC TX descriptor; the key information can be packet size, memory address, checksum offload setting, and VLAN configuration. PMD is responsible for every packet to be sent; the NIC (HW) TX descriptor is set up so that NIC can execute the TX action with configuration. A packet may occupy one or more mbuf elements. PMD needs to identify an end-of-packet (EOP) flag, which is to notify that NIC is the last mbuf of a complete packet.

At TX side, if PMD is working on busy poll, PMD will need to know if the packet TX is done by NIC. "RS" bit is such a flag to inform NIC; then NIC will write back the TX status, and it might be implementation specific. To achieve a better efficiency and performance, the driver may not require every packet to update the TX status; indeed, this depends on NIC implementation, and some NICs require every single packet to report the TX status at a time. Some NICs can support the status report for multiple packets. The driver implementation is responsible for this specific detail. If NIC writes back the packet TX status, PMD is responsible for learning it and taking actions. For example, the NIC TX descriptor will be cleared and mbuf will be recycled for future use. The write-back may be different, such as descriptor-based write-back or header-based write-back.

Busy poll is executed by the assigned core on the specified queues; the core frequency control is out of DPDK scope, but the software can be intuitive to look at the incoming packet rates (network telemetry) at a certain period; it can give a hint on if the core frequency can be turned up or down accordingly. The good enough clock frequency can sustain the packet forwarding without loss; the core will still be occupied (not for other tasks), but it can reduce the power consumption, which is very useful as energy-saving design.

A typical Ethernet device has the below data structure, and the device driver (PMD) needs to follow this framework:

```
struct rte_eth_dev {
    eth_rx_burst_t rx_pkt_burst; /* Pointer to PMD function. */
    eth_tx_burst_t tx_pkt_burst; /* Pointer to PMD function. */
    eth_tx_prep_t tx_pkt_prepare; /* Pointer to PMD function. */
```

```
/**
 * Next two fields are per-device data but *data is shared
 * between primary and secondary processes
 */
  struct rte_eth_dev_data *data; /* Pointer to device data. */
  void *process_private; /* Pointer to per-process data. */
  const struct eth_dev_ops *dev_ops; /* PMD Functions */
  struct rte_device *device; /**< Backing device */
  struct rte_intr_handle *intr_handle;
 /** User application callbacks for NIC interrupts */

  struct rte_eth_dev_cb_list link_intr_cbs;
  /* User-supplied functions called from rx_burst to
  * post-process received packets before passing to user*/

  struct rte_eth_rxtx_callback
       *post_rx_burst_cbs [RTE_MAX_QUEUES_PER_PORT];

  /**
  * User-supplied functions called from tx_burst to
  * pre-process received packets before passing to the driver
  */
 struct rte_eth_rxtx_callback
     *pre_tx_burst_cbs[RTE_MAX_QUEUES_PER_PORT];

 enum rte_eth_dev_state state; /* Flags of port state */

 void *security_ctx; /* Context for security ops */

} __rte_cache_aligned;
```

7.1.3 HYBRID MODE

Internet traffic has the tidal effects; sometimes, it beats with human activities; the network traffic may be very low at certain times. For example, after midnight, a low-load scenario will occur on the high-speed network system at some locations. If the busy poll mode is enabled, the core can run at full speed. In reality, this may waste the computing cycles and it doesn't save energy and money. Therefore, the hybrid of interrupt and poll modes is also introduced in DPDK.

HIP (hybrid interrupt poll) is given in DPDK example "**l3fwd-power**". Its design framework is described below (Figure 7.1):

1. The DPDK application creates a polling thread, which runs in the busy poll mode to check if any packets need to be received. By default, the interrupt is turned off.
2. The polling thread detects the incoming packets on a core basis. If no packet is received, a conservative approach is implemented to decide the core idle time and then what's next step. Two actions are given in case of no packets are received in the example.

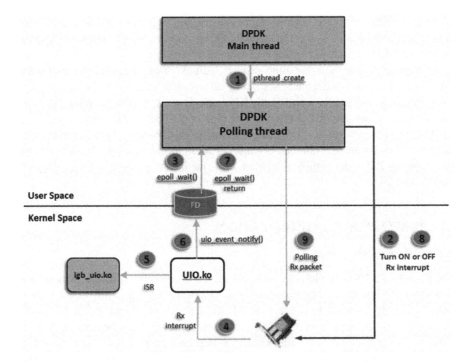

FIGURE 7.1   Hybrid model (poll and interrupt).

- Within the core idle threshold (configurable by MACRO: SUSPEND _ THRESHOLD), use "Pause" instruction to delay the busy poll for a specified time, which is implemented by "rte _ delay _ us" API.
- If the idle exceeds the configured threshold, interrupt/epoll mode is supported/enabled at the port/queue level (rte _ eth _ dev _ rx _ intr _ ctl _ q), the example strategy will turn on the port/queue interrupt, and the core enables one-time interrupt service before sleep.

3. The core (polling thread) will block on epoll event: "rte _ epoll _ wait" is used here, which depends on epoll_wait.
4. If a packet comes on the wire, the NIC will trigger a "Rx Interrupt".
5. Depending on the kernel module, it will notify the igb_uio module if "UIO" is in place.
6. UIO (user space I/O) will notify the user space with "uio _ event _ notify".
7. The sleep core is waked up, and it will return on "rte _ poll _ wait".
8. The polling thread is back to work, and it will disable the port/queue interrupt and continue the busy poll mode.

The aforementioned example decides when to use the poll mode and when to use the interrupt mode. It demonstrates a conservative approach before using interrupt mode. If core enters the sleep mode, there is increased latency to get back to the poll mode; if large traffic arrives at the sleep mode, appropriate queue setup with a large number of mbuf elements may help to avoid the packet loss.

In order to utilize the computing capacity of the processor more efficiently, the user needs to decide what's the best strategy on poll, interrupt, or core frequency up/down decisions on the busy poll. The HIP mode depends on the interrupt notification from the kernel modules, which can be UIO or VFIO (virtual function I/O). For VFIO-based implementation, the interrupt is supported at the queue level; every rxq can have its own interrupt number, because VFIO supports multiple MSI-X interrupt numbers. For UIO-based implementation, only one interrupt number is shared by all the RX queues.

7.2 PERFORMANCE OPTIMIZATION

7.2.1 BURST PROCESSING

The packet RX/TX is a relatively straightforward computation process, which consists of mbuf allocation/release, and packet descriptor read and write; there are multiple read/write accesses to the memory. If the data access happened in the cache (cache hit), it is good for access latency reduction. Otherwise, cache miss could lead to the actual memory access with a high latency impact, and hence, it is a performance killer. CPU caches are designed in multiple levels; each level has a different size and access latency. In the Linux system, "dmidecode memory" can help to get the cache info (Figure 7.2).

The aforementioned example shows three levels of the CPU cache; each level has a different size and speed. For example, "Socket Designation: L1-Cache" refers to the first-level cache and "Maximum Size: 1152 KB" indicates that the total size is 1152 K bytes.

How to write software to make the best use of the cache? DPDK uses the burst processing, and it is a well-known optimization idea. It decomposes the packet RX/TX into multiple small processing stages, allowing the adjacent data access and similar processing together. Thus, this can minimize the memory access for the multiple data read/writes, and the cache utilization can be highly effective. Packet burst processing is about to handle multiple packets (such as 8, 16, or even 32) at a time. The number of packets is determined by the DPDK user.

The NIC descriptor (RX or TX) is generally 16 or 32 bytes long (e.g., Intel® 82599, X710). x86 cache line size is 64 bytes. Each cache line can hold two or four "receive" descriptors. The NIC can combine the descriptor update (such as four or eight) together, and it is still cache line-aligned. The prefetch mechanism (hardware or software based) can help the adjacent cache lines to be accessed at one time. If four cache lines are updated, 16 descriptor (each size is 16 bytes long) accesses can be combined. If each memory access only updates one descriptor, the access efficiency is low because the optimization opportunity is lost to combine with 15 NIC descriptors.

The burst processing use is a pattern widely employed in DPDK. The user can decide how many packets are required for packet burst operation. Here are the function interfaces.

```
Handle 0x0018, DMI type 7, 19 bytes
Cache Information
        Socket Designation: L1-Cache
        Configuration: Enabled, Not Socketed, Level 1
        Operational Mode: Write Back
        Location: Internal
        Installed Size: 1152 kB
        Maximum Size: 1152 kB
        Supported SRAM Types:
                Synchronous
        Installed SRAM Type: Synchronous
        Speed: Unknown
        Error Correction Type: Single-bit ECC
        System Type: Instruction
        Associativity: 8-way Set-associative

Handle 0x0019, DMI type 7, 19 bytes
Cache Information
        Socket Designation: L2-Cache
        Configuration: Enabled, Not Socketed, Level 2
        Operational Mode: Varies With Memory Address
        Location: Internal
        Installed Size: 4608 kB
        Maximum Size: 4608 kB
        Supported SRAM Types:
                Synchronous
        Installed SRAM Type: Synchronous
        Speed: Unknown
        Error Correction Type: Single-bit ECC
        System Type: Unified
        Associativity: 8-way Set-associative

Handle 0x001A, DMI type 7, 19 bytes
Cache Information
        Socket Designation: L3-Cache
        Configuration: Enabled, Not Socketed, Level 3
        Operational Mode: Varies With Memory Address
        Location: Internal
        Installed Size: 46080 kB
        Maximum Size: 46080 kB
        Supported SRAM Types:
                Synchronous
        Installed SRAM Type: Synchronous
        Speed: Unknown
        Error Correction Type: Single-bit ECC
        System Type: Unified
        Associativity: Fully Associative
```

FIGURE 7.2   Cache information.

```
static inline uint16_t rte_eth_rx_burst(uint8_t port_id,
uint16_t queue_id, struct rte_mbuf **rx_pkts, const uint16_t
nb_pkts);

static inline uint16_t rte_eth_tx_burst(uint8_t port_id,
uint16_t queue_id, struct rte_mbuf **tx_pkts, uint16_t
nb_pkts);
```

nb _ pkts, the last parameter, specifies the total packet in burst each time. When it is set to 1, only one packet is sent or received at a time. The default number can be 32

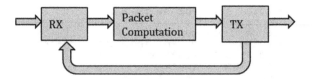

FIGURE 7.3   A simplified packet RX/TX example.

in many examples. Figure 7.3 shows the simplest packet sending/receiving process. After a packet is received, the application may perform a lot of packet-based computation before the packet TX.

In the non-burst design, only one packet was received at one time, and then it is processed and sent, at an individual basis. For each packet RX, the packet memory is loaded into cache, the adjacent data is also loaded to the cache, and the memory access always happened with a cache line size (like 64 bytes). The prior data in the cache can be vacated while doing the packet computation; then the next use of the adjacent data becomes a problem; it may have to load the data from a lower-level cache or access the memory again.

In the packet burst design, multiple packets can be processed at a time. When the data is loaded in the first time (for the first packet), the adjacent data may be useful to the next packet's use. This means only one memory access (or lower-level cache access) is required to receive two or more packets. This way, fewer memory access (or low-level cache access) is required to complete the multiple packets of RX. Overall, this reduces the average packet receiving cycle and improves the overall system performance.

If one packet is sent one at a time, the memory access leads to the cache update in units of a cache line, and data can be loaded to the cache, but not used at all. This is not an optimal way of using the critical cache space. The packet burst processing allows multiple packets to be sent at a time. It uses less memory access (or lower-level cache) synchronization while sending more packets. Overall, this improves the performance.

7.2.2   BATCH PROCESSING

When studying instruction optimization, latency and throughput are two known concepts. Both describe the instruction overhead in a context of CPU's parallel execution units.

- *Latency*: The number of clock cycles to run an instruction.
- *Throughput*: The number of clock cycles to run the same instruction again.

Latency describes the waiting interval between two instructions, whereas throughput describes the concurrency capacity. The instruction latency is relatively fixed, concurrency can enhance the overall system performance, and issuing more instructions will help to increase concurrency. A simple example can help to illustrate the two basic concepts in Figure 7.4.

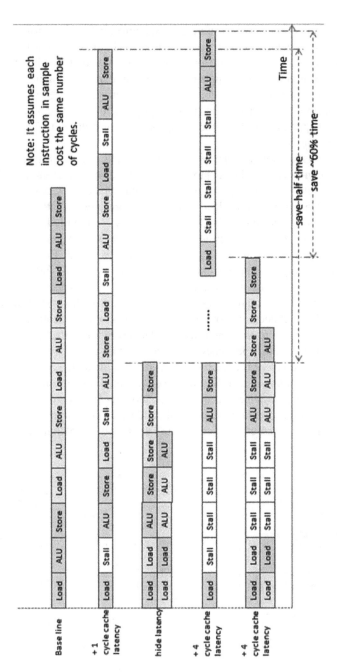

FIGURE 7.4 Latency hiding example.

To simplify the model, let's assume a transaction only has three instructions (load, ALU, and store); each instruction can start at the end of the previous instruction; the latency and the throughput is just one clock cycle. The first line is a total time to execute four transactions on a single execution unit, thus resulting in 12 clock cycles in total. All transactions are independent of each other. If there are two execution units on load and ALU instructions, the throughput is reduced to 0.5 cycles (allowing two loads to happen in parallel), and the spent cycles is reduced to 7 clock cycles, as shown in the line 3.

A more realistic example is a memory access, and its cost has instruction overhead and access latency. Even if the data to be accessed is already present in the L1 cache, 3 to 4 clock cycles are still required to use the data. It takes 4 clock cycles for "load" instruction. If all instructions (load, ALU, and store) are executed in a sequential order, as shown in the fourth line, 28 clock cycles are required. It takes 4 cycles ("stall") until the "load" instruction is completed. As shown in the fifth line, each "load" takes 4 cycles until data can be used; issuing two "load" instructions are possible with the multiple data paths; now it only takes 3 cycles in "stall" (instead of 4). This demonstrates a latency hiding technique of the "load" instruction. Conceptually, the independent and repeatable transactions can use this method to hide more latency.

Modern CPU architecture is optimized with out-of-order execution and high concurrency. The effective method on instruction latency hiding is the batch processing; it is valid if there is no data dependency. The repetitive task can be done with a successive loop. For a more complex task, the compiler is responsible for the code generation, and the compiler cannot optimize a large iterative sequential instruction with out-of-order execution. Software developer has the optimization opportunity to break the complex computation into multiple tasks sequentially and perform each in a batch processing approach. This is a common practice used in DPDK code development (Figure 7.5).

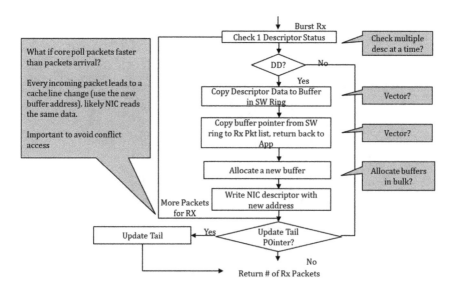

FIGURE 7.5 Packet RX and optimization opportunity.

Here is an example of packet RX. The main procedure includes the following:

- Checking the descriptor/DD flag;
- Processing packet descriptors;
- Allocating new buffers;
- Refilling the NIC descriptor with new buffer address;
- Updating the tail pointer of NIC descriptor ring.

As discussed earlier, packet RX needs the multiple memory access (such as descriptor and buffer read/write); at the instruction level, memory accesses are translated into the memory load/store, a big portion of the entire processing cycles. Most steps (except for updating the tail pointer) can be implemented with packet batching approach. And the tail pointer update is not necessary for every individual packet, and the update is essentially MMIO (memory-mapped I/O)-based access, which is expensive operation. Reducing the MMIO access can save cycles; one option is to update the tail pointer (once) for all received packets. Overall, the batching processing is useful and effective here.

In theory, the CPU and NIC (DMA engine) may try to access the same data. There are two kinds of data access:

- *Read–write conflict*: Use the DD flag as an example. If packet RX is completed at NIC side, NIC will write the flag. Meanwhile, the CPU is polling this flag periodically.
- *Write–write conflict*: CPU needs to assign the new packet buffer and refills them to NIC RX descriptor ring. Meanwhile, NIC will write the ring as part of packet RX, which is a DMA write-back.

x86 cache design supports the data coherency. When both CPU and NIC try to access the same data, it leads to additional synchronization overhead, in particular the scenario of "write access". When NIC writes the data, it is often done on a cache line basis; the below example avoids the write–write conflict, and it is important to control the steps of updating the tail pointer (Figure 7.6).

The below procedure is a new packet RX processing, which is different as it is based on the batch processing method. Intel® 82599 NIC defines its descriptor size as 16 bytes long; four descriptors' access can be combined together as one cache line. The NIC always tends to write data in a cache line size anyway. Combining four packets in a batch saves cycles. Allocating the new packet buffer, refilling the NIC descriptors, and updating the descriptor ring are the large proportion of the packet RX. These procedures can be separated from other tasks such as checking the DD flag and writing the packet descriptors. The entire processing cycle can be shortened with a few deliberate techniques, which are as follows:

- Batch processing to hide the access latency;
- Batch processing on effective cache access;
- Timing the data access to avoid "core/NIC" conflict.

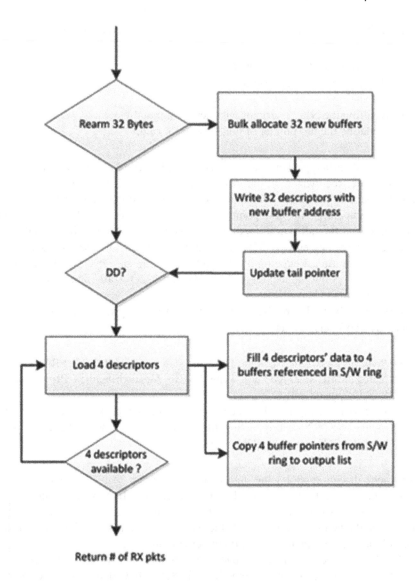

FIGURE 7.6 Batch processing design.

In DPDK, ixgbe PMD has the batch processing for the packet RX/TX. The PMD checks if 8 packets are ready on the NIC's receive queue. If so, the software will parse the hardware descriptor and translate the NIC descriptor to software-based packet metadata fields (mbuf). This is repeated until 32 packets are received, or no further packets arrived. Then, PMD will check if there is a need to allocate more packet buffers. If the threshold is reached, PMD will ask for many mbufs at a time. The final step will look into the actual number of packets in a burst, and refill the packet buffers as needed.

The packet receiving process is broken down into three relatively small tasks. Each task is aimed to handle many packets (e.g., between 8 and 32 packets). The bulk interface provides the optimization opportunity. Compared to the traditional approach, software interface handles one packet at a time, the batched processing is more effective for a large number of packet processing, and the software design takes the CPU instruction and memory access principles into account.

The batching method is not always used, but it is used when certain optimization conditions are met. Here is one example in the ixgbe PMD, which explains when it is not recommended to use.

```
/*
 * Make sure the following pre-conditions are satisfied:
 * rxq->rx_free_thresh >= RTE_PMD_IXGBE_RX_MAX_BURST
 * rxq->rx_free_thresh < rxq->nb_rx_desc
 * (rxq->nb_rx_desc % rxq->rx_free_thresh) == 0
 * rxq->nb_rx_desc<(IXGBE_MAX_RING_DESC- RTE_PMD_IXGBE_
   RX_MAX_BURST)
 * Scattered packets are not supported. This should be checked
 * outside of this function.
 */
```

7.2.3 SIMD

Batch processing is a big step to achieve performance optimization. Is there more room available? From the data access perspective, Table 7.1 describes the speed and cost in different Intel® processors. Nehalem, SNB (Sandy Bridge), and HSW (Haswell) are used for Intel® Xeon processor as code name in different generations. Data access throughput is more than doubled from Nehalem to HSW. The packet processing is a heavy user of the memory access. The effective use of cache/memory bandwidth and throughput is the most important area to be studied (Table 7.1).

TABLE 7.1

Cache and Data Bandwidth on CPUs

Metric\Intel® CPU code name	Nehalem	Sandy Bridge	Haswell	Comments
L1 Instruction Cache	32K	32K	32K	
L1 Data Cache	32K	32K	32K	
Hit Latency	4/5/7	4/5/7	4/5/7	
Bandwidth (bytes/ cycles)	16+16	32+16	64+32	2 loads+1 store
L2 Cache	256K	256K	256K	
Hit Latency	10	12	12	
Bandwidth (bytes/ cycles)	32	32	64	

For the read access (load latency is 1 cycle and throughput is 0.5 cycles) in Intel® 64-bit processor, the standard register size is just 8 bytes (64 bits); 16 bytes of data can be loaded in one cycle to the CPU registers. However, in theory, SNB can load 32 bytes per cycle; only half of the bandwidth is used on the standard registers. SIMD instructions can make a more effective use of bandwidth.

Per its name, SIMD is single-instruction and multiple data access. The Intel® SSE instruction is based on 128-bit register, whereas AVX (advanced vector extensions) provides a 256-bit register and AVX512 provides a 512-bit register. From the register width perspective, SSE register can make full use of the SNB's bandwidth; while HSW is designed with AVX register, this is the optimal use from the data bandwidth-only perspective. Similar theory applies to AVX512 registers on Skylake and Cascade lake processors today.

The idea is to use the SSE/AVX registers in PMD, the theory works, and the software implementation is a new challenge. DPDK has implemented the "vectorized PMD" for ixgbe (Intel® Ethernet product series). SIMD accesses the wide registers, extra care is required when defining the data structure, and its members' access needs to gain performance benefits. The improper data layout may lead to performance degradation with SIMD. "rte_mbuf" is an important data structure, and its design has the consideration to be SIMD-friendly. Not all PMD supported the vectorized PMD, which uses the SIMD instruction heavily. For a code deep dive, please look into the source file at:

```
/source/drivers/net/ixgbe/ixgbe_rxtx_vec_sse.c (for x86);
/source/drivers/net/ixgbe/ixgbe_rxtx_vec_neon.c (for ARM);
```

7.3   PLATFORM CONFIGURATION

DPDK is famous for the performance optimization; the system configuration (hardware and software) has a big impact. In essence, the hardware system determines the best possible performance, which consists of CPU, chipset, I/O interface, memory, and NICs; all of them are configurable. For example, where to insert a NIC if there are multiple slots available? Software configuration is equally critical, and this determines if all hardware potential can be realized; some examples are firmware, operating system and kernel, compiler versions, and configurations.

7.3.1   HARDWARE IMPACTS

Take Intel® Xeon server system as an example; let's visit the processor, memory, and I/O units. CPU frequency has a linear impact on the performance. Different CPUs may support different sets of instructions. In general, new processor provides a better performance and new instructions. Some recent examples are AVX instruction sets. DPDK developers have utilized the new instructions to write software; some specific modules are vectorized PMD, ACL, and rte_memcpy. Another example is the AES-NI for crypto processing. Another processor change is increasing cache size and efficiency. On a dual-socket server system, performance drops if the data access needs to be made across the socket. In a Linux system, the command "cat/proc/cpuinfo" can show the processor details (Figure 7.7).

```
processor       : 63
vendor_id       : GenuineIntel
cpu family      : 6
model           : 63
model name      : Intel(R) Xeon(R) CPU E5-2699 v3 @ 2.30GHz
stepping        : 2
microcode       : 0x1d
cpu MHz         : 2300.000
cache size      : 46080 KB
physical id     : 1
siblings        : 28
core id         : 16
cpu cores       : 18
apicid          : 97
initial apicid  : 97
fpu             : yes
fpu_exception   : yes
cpuid level     : 15
wp              : yes
flags           : fpu vme de pse tsc msr pae mce cx8 apic sep mtrr pge mca cmov pat pse36
 clflush dts acpi mmx fxsr sse sse2 ss ht tm pbe syscall nx pdpe1gb rdtscp lm constant_tsc
 arch_perfmon pebs bts rep_good nopl xtopology nonstop_tsc aperfmperf eagerfpu pni pclmulqd
 q dtes64 monitor ds_cpl vmx smx est tm2 ssse3 fma cx16 xtpr pdcm pcid dca sse4_1 sse4_2 x2
 apic movbe popcnt tsc_deadline_timer aes xsave avx f16c rdrand lahf_lm abm arat epb pln pt
 s dtherm tpr_shadow vnmi flexpriority ept vpid fsgsbase tsc_adjust bmi1 hle avx2 smep bmi2
 erms invpcid rtm xsaveopt
bugs            :
bogomips        : 4593.49
clflush size    : 64
cache_alignment : 64
address sizes   : 46 bits physical, 48 bits virtual
power management:
```

FIGURE 7.7 The processor information.

The memory controller, RAM, and its physical layout affect the system performance. In general, the memory controller on the server platform supports multiple memory channels, and DPDK expects the physical RAMs are inserted evenly to all memory channels, which helps the optimal concurrent access to the memory and helps for the large packet-processing load. In a Linux system, "dmidecode memory" can help to inspect the memory details (Figure 7.8).

* "Size:8192 MB" in Figure 7.9 shows the RAM size.
* "Locator:DIMM_E1" shows the physical location of the RAM.
* "Bank Locator:CPU 1" indicates that RAM is connected with CPU 1.
* "Speed:1067 MHz" shows the memory frequency.

```
Handle 0x002E, DMI type 17, 34 bytes
Memory Device
        Array Handle: 0x001C
        Error Information Handle: Not Provided
        Total Width: 72 bits
        Data Width: 72 bits
        Size: 8192 MB
        Form Factor: DIMM
        Set: None
        Locator: DIMM_E1
        Bank Locator: CPU 1
        Type: <OUT OF SPEC>
        Type Detail: Synchronous
        Speed: 1067 MHz
        Manufacturer: 0x44
        Serial Number: Unknown
        Asset Tag: Unknown
        Part Number: Unknown
        Rank: 1
        Configured Clock Speed: Unknown
```

FIGURE 7.8 The memory information.

As mentioned repeatedly, the access latency to the local or remote memory is very different, remote access latency may add more than 50% penalty, or even double, and it varies with respect to the different processors. From the software perspective, extra attention is required on memory use. DPDK is designed to be local or remote socket aware, and many software APIs are asking "socket" as the explicit input parameter.

PCIe is I/O interface on the server platform. NIC is plugged into PCIe slot; it is socket relevant in a multisocket system. PCIe speed is negotiable between NIC and platform. Ideally, the local core drives the local PCIe slot. Using the remote core to handle the PCIe card will lead to extra inter-CPU communication, thereby slowing down the performance. PCIe slot is always attached to one CPU, and this is finalized in the motherboard design. Balanced I/O design is important for Telecom server, which means that I/O cards (like NIC) can be plugged in a balanced way, so that the dual socket can have the balanced packet processing for each socket.

Software developer can check the server hardware manual to determine the PCIe slot and CPU connectivity.

To understand the NIC location on socket, it can check the PCIe device's BFD info:

- If it is less than 80 (e.g., 0000: 03: 00.0), the NIC is attached to CPU0.
- If it is equal to 80 or more (e.g., 0000: 81: 00.0), it is attached to CPU1.

However, the software method may subject to change, and it depends on OS kernel and driver configuration.

Figure 7.9 describes a typical board layout on a dual-socket server; there are six PCIe slots available; each slot lane is finalized in the motherboard design; and the connection to the socket is clear as well. Intel® Fortville NIC supports 40 Gbe I/O with PCIe Gen3x8, and it should be inserted into a physical slot, which can support PCIe Gen3x8. On a Linux platform, the use of the command "lspci–vvv" can inspect the PCIe interface and details. PCIe "Extended Tag" feature needs to be highlighted, and it has a big performance impact on the Intel® 40 Gbe NIC. It might be disabled in some default platform settings. Here are a few options to enable it:

- Configure BIOS (if available).
- Enable it via the command "setpci" in a Linux system.
- Enable it via the software function provided by DPDK (Figure 7.10).

NIC is the most important networking device; it decides the physical I/O interface of the system; some typical NIC interfaces are listed as follows:

- 2 × 10/25/50 Gbe;
- 4 × 10/25 Gbe;
- 1 × 10/25/50/100 Gbe;
- 2 × 10/25/50/100 Gbe.

I/O interfaces do not guarantee the NIC's maximum networking performance. Don't assume that 2x40Gbe Ethernet ports will deliver 80 Gbps I/O simultaneously. Other

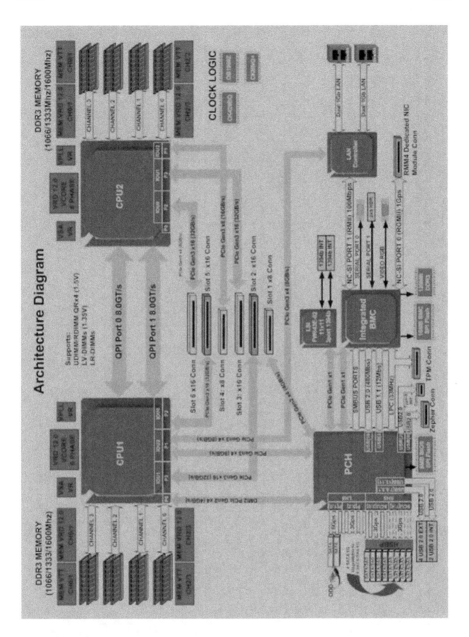

FIGURE 7.9 A dual-socket server and PCIe slots.

system factors such as the Ethernet chip capability and PCIe slot (speed and lanes) affect the I/O performance. The same Ethernet chip can be assembled into NIC by the different vendors. The NIC can be sold under different brand names. Its I/O performance may vary greatly.

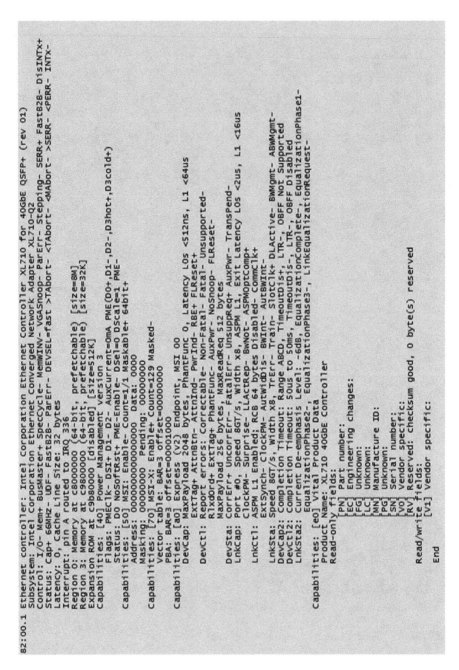

FIGURE 7.10 PCIe device information.

In some rare scenarios, a PCIe Gen3-capable device may be downgraded to PCIe Gen2 due to device/platform negotiation. Here is an example. Intel® NIC XL710 is shipped with on-board firmware. The firmware version can contribute to PCIe speed downgrade. If an unexpected performance downgrade happened, it might be caused by the early firmware version. In general, always try to ensure that NIC has the latest firmware.

Most NICs generally support multi-queues for packet I/O. When receive-side scaling (RSS) is enabled, NIC can distribute the packets into multiple queues, and then multiple cores can handle one or more queues for load balancing. Depending on the exact NIC, the line rate of Ethernet port may not be achieved with a single queue, and the line rate may be achieved using the multiple queues.

7.3.2   Software Impacts

In general, high performance and energy saving might be contradictory. In order to achieve the best performance, the firmware setting is important. For example, turn off the power-saving modes of the CPU and devices, and ensure that the memory can run at the highest frequency.

Generally, a new but stable kernel version is preferred for Linux OS. Different kernel versions hint the different features. DPDK depends on the HugePage (2 MB or 1 GB) mechanism to reduce the performance impact of TLB (translation lookaside buffer) miss. In order to use a hugepage with a size = 1 GB, the Linux kernel boot parameter can be configured in the following.

```
'default_hugepagesz = 1G hugepagesz = 1G hugepages = 8'.
```

In a Linux system, "cat /proc/meminfo" can help to look at the memory info such as huge page allocation and actual use (Figure 7.11).

DPDK model is to assign the dedicated cores to run packet RX/TX tasks, and it will try to avoid Linux OS to use the assigned cores. Reserving the core resource can be done at the Linux kernel boot: "isolcpus = 2,3,4,5,6,7,8". This prevents the specified cores from being scheduled by Linux.

Intel® XL710 NIC needs to be turn on the PCIe "Extended Tag" to make sure that PCIe bus is fully utilized, the "Extended Tag" is 8 bits wide, the default setting only uses 5 bits effectively, and finally it supports 32 transactions at a time. When all 8 bits is turned on, 256 transactions can be issued at a time. This affects the high-speed port (40Gbe and beyond). Turning on the below DPDK compilation flags will help.

```
CONFIG_RTE_PCI_CONFIG=y
CONFIG_RTE_PCI_EXTENDED_TAG="on"
```

In the dual-socket system, the principle is to use the local core to drive the local Ethernet port. User needs to know this rule is not broken. On a Linux system, "lscpu" can check the core ID of the processor (Figure 7.12).

"NUMA node0 CPU(s): 0–17, 36–53" indicates that the cores 0–17 and 36–53 belong to Node (Socket) 0, while others belong to Node 1. Linux command

```
MemTotal:              65918896 kB
MemFree:               56019272 kB
MemAvailable:          56854160 kB
Buffers:                 103532 kB
Cached:                  784660 kB
SwapCached:                   0 kB
Active:                  620876 kB
Inactive:                480316 kB
Active(anon):            222420 kB
Inactive(anon):            3188 kB
Active(file):            398456 kB
Inactive(file):          477128 kB
Unevictable:              15792 kB
Mlocked:                  15792 kB
SwapTotal:             16457724 kB
SwapFree:              16457724 kB
Dirty:                        0 kB
Writeback:                    0 kB
AnonPages:               228896 kB
Mapped:                  226616 kB
Shmem:                    10040 kB
Slab:                    187032 kB
SReclaimable:             70252 kB
SUnreclaim:              116780 kB
KernelStack:              10640 kB
PageTables:               18432 kB
NFS_Unstable:                 0 kB
Bounce:                       0 kB
WritebackTmp:                 0 kB
CommitLimit:           45222868 kB
Committed_AS:           1407488 kB
VmallocTotal:       34359738367 kB
VmallocUsed:             474728 kB
VmallocChunk:       34325068564 kB
HugePages_Total:              8
HugePages_Free:               0
HugePages_Rsvd:               0
HugePages_Surp:               0
Hugepagesize:           1048576 kB
DirectMap4k:              13160 kB
DirectMap2M:            1964032 kB
DirectMap1G:           67108864 kB
```

FIGURE 7.11 System memory information.

"lspci-NN | grep Eth" can list the NIC devices and its PCI device info (Figure 7.13).

"82:00.0 Ethernet controller [0200]: Intel® Corporation Ethernet controller XL710 for 40GbE QSFP+" shows PCIe device "0000:82:00.0" is an Intel® XL710 NIC on Node (Socket) 1.

Intel® Fortville 40G NIC (XL710) supports 2x40Gbe Ethernet port; however, its PCIe interface is designed as Gen3 x8, which cannot sustain 2x 40 Gbps forwarding data rate. To build a networking system with 80 Gbps I/O, two such NICs are required. If only a single queue is used in this NIC, the line rate speed of small packet (e.g., 128 bytes) is not achievable, so two or more queues are necessary to

```
Architecture:            x86_64
CPU op-mode(s):          32-bit, 64-bit
Byte Order:              Little Endian
CPU(s):                  64
On-line CPU(s) list:     0-63
Thread(s) per core:      1
Core(s) per socket:      18
Socket(s):               2
NUMA node(s):            2
Vendor ID:               GenuineIntel
CPU family:              6
Model:                   63
Model name:              Intel(R) Xeon(R) CPU E5-2699 v3 @ 2.30GHz
Stepping:                2
CPU MHz:                 2300.000
CPU max MHz:             2300.0000
CPU min MHz:             1200.0000
BogoMIPS:                4593.49
Virtualization:          VT-x
L1d cache:               32K
L1i cache:               32K
L2 cache:                256K
L3 cache:                46080K
NUMA node0 CPU(s):       0-17,36-53
NUMA node1 CPU(s):       18-35,54-63
```

FIGURE 7.12   CPU information.

```
03:00.0 Ethernet controller [0200]: Intel Corporation Ethernet Controller 10-Gigabit X540-AT2 [8086:1528] (rev 01)
03:00.1 Ethernet controller [0200]: Intel Corporation Ethernet Controller 10-Gigabit X540-AT2 [8086:1528] (rev 01)
05:00.0 Ethernet controller [0200]: Intel Corporation Ethernet Controller X710 for 10GbE SFP+ [8086:1572] (rev 01)
05:00.1 Ethernet controller [0200]: Intel Corporation Ethernet Controller X710 for 10GbE SFP+ [8086:1572] (rev 01)
05:00.2 Ethernet controller [0200]: Intel Corporation Ethernet Controller X710 for 10GbE SFP+ [8086:1572] (rev 01)
05:00.3 Ethernet controller [0200]: Intel Corporation Ethernet Controller X710 for 10GbE SFP+ [8086:1572] (rev 01)
81:00.0 Ethernet controller [0200]: Broadcom Corporation NetXtreme BCM5751 Gigabit Ethernet PCI Express [14e4:1677] (rev 01)
82:00.0 Ethernet controller [0200]: Intel Corporation Ethernet Controller XL710 for 40GbE QSFP+ [8086:1583] (rev 01)
82:00.1 Ethernet controller [0200]: Intel Corporation Ethernet Controller XL710 for 40GbE QSFP+ [8086:1583] (rev 01)
84:00.0 Ethernet controller [0200]: Intel Corporation Ethernet Controller X710 for 10GbE SFP+ [8086:1572] (rev 01)
84:00.1 Ethernet controller [0200]: Intel Corporation Ethernet Controller X710 for 10GbE SFP+ [8086:1572] (rev 01)
86:00.0 Ethernet controller [0200]: Intel Corporation Ethernet Controller XL710 for 40GbE QSFP+ [8086:1583] (rev 01)
86:00.1 Ethernet controller [0200]: Intel Corporation Ethernet Controller XL710 for 40GbE QSFP+ [8086:1583] (rev 01)
```

FIGURE 7.13   NIC and PCIe information.

make sure that NIC works at the highest speed. To make sure that the system has a balanced workload, one NIC is plugged to Socket 0, whereas another goes to Socket 1. Let's assume that their PCI devices are known as 0000:82:00.0 and 0000:85:00.0, four queues are used for each Ethernet port, l3fwd example is used for forwarding, eight cores from Node 1 can be assigned (such as core 20, 21, 22, 23, 24, 25, 26, and 27), and the example configuration parameter pattern (port _ id, queue _ id, core _ id) can be:

```
'./l3fwd -c 0xff00000-n 4-w 82:00.0-w 85:00.0-- -p 0x3 -
config'(0,0,20), (0,1,21), (0,2,22), (0,3,23), (1,0,24),
(1,1,25), (1,2,26), (1,3,27)''.
```

With this configuration, each port is configured with four queues, and a dedicated core is assigned to the queue. In the ideal world, each core shares the load evenly. NIC can use "RSS" feature to distribute packets to four queues as evenly as possible; in a test environment, incoming traffic is not predictable, so the actual load may not be evenly distributed. To simulate an even incoming traffic scenario, a simple

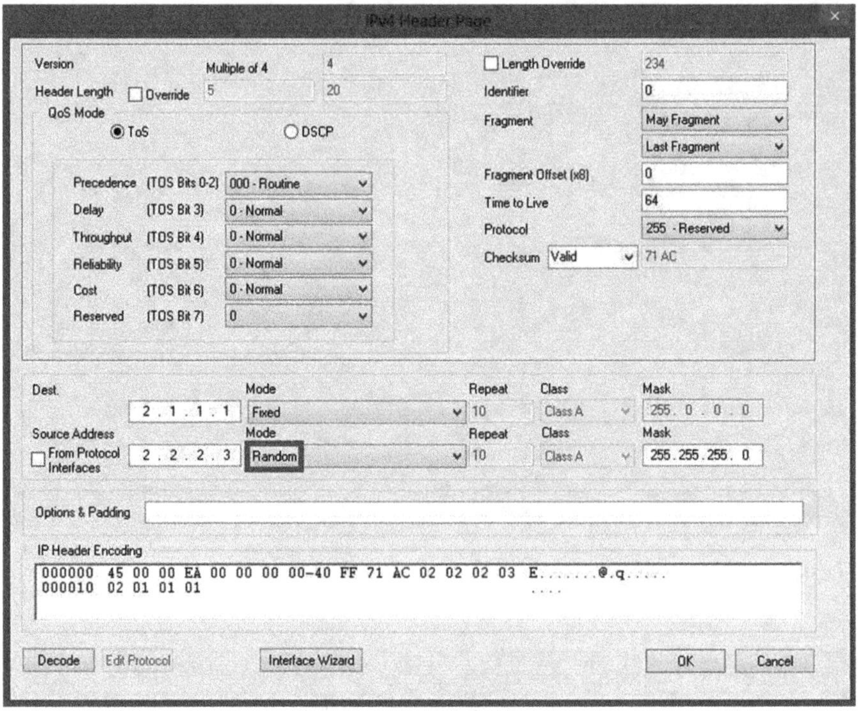

FIGURE 7.14   IXIA configuration: Create random source IP.

method is to configure the source IP address as random, and this can be done by the hardware test equipment like IXIA (or Spirent) (Figure 7.14).

If the traffic is not evenly distributed to the queues, some cores will be underutilized if the incoming traffic is low on that queue. RSS algorithm and configuration are another NIC tuning.

7.4   NIC TUNING

7.4.1   RXQ/TXQ Length

The length of the receive queue refers to the number of receive descriptors in the queue. Each receive descriptor is associated with mbuf. The queue length indicates how many packets that NIC RX can hold before PMD comes to get packets. The longer length means more packets can be held by NIC, and it also uses more memory in advance. NIC hardware defines the maximum queue length. If the queue length is set short, less mbuf shall be distributed to each queue, and some queue length is set as 128 by default. If there is a packet loss in the high traffic scenario, maybe adjust it to 512 or 1024. A similar configuration can be done on the length of transmit queue.

7.4.2   RXQ/TXQ Threshold

While a packet is received, its receive descriptor is released, the tail update does not happen for each single packet, and DPDK wants to reduce MMIO access. The update is performed only when a certain threshold value, or rx _ free _ thresh, is reached. This threshold is set as 32 descriptors by PMD. It can be adjusted.

At the TX side, it is a similar idea, i.e., tx _ free _ thresh is set as 32 by PMD. Users can set different threshold values. A small threshold may lead to frequent MMIO access with less performance, whereas a large threshold may result in the packet loss and increased latency, as the descriptor is not used as fast as possible.

7.4.3   Conclusion

DPDK is very effective in driving the high-speed NIC. The software secret recipe is discussed here. In addition, the platform configuration and tuning need to look at both hardware and software components. The key concepts are related to the platform layout, memory and I/O interface, and NIC setting. Chapter 8 will go a little further on NIC internals.

8 NIC-Based Parallellism

Jingjing Wu, Xiaolong Ye, and Heqing Zhu
Intel®

CONTENTS

Multi-queue NIC supports multiple queues in both RX and TX directions, and this hardware feature is a commonly available and widely used feature, the incoming packets can be moved into different queues, and it is a load balancing mechanism to receive large amounts of traffic. Flow classification is a way to identify the incoming packets going into multiple queues. When combining them, NIC provides a scalable I/O mechanism, and when associating the NIC multi-queue with CPU multi-core, this creates the hardware foundation to build the multi-threading packet processing architecture, and this can build a high-performing networking system with a server platform (Figure 8.1).

Intel® NICs are used as a reference to describe how its multi-queue and flow classification work and how DPDK supports them. In general, standard traditional NIC (such as Intel® 82599) is designed for enterprise and data center servers, and some NICs may not support the telecom networking protocols such as PPPoE, GTP-U, Segment Routing, and MPLS/GTP. The later NICs (such as Intel® XL710) can be extended to support the telecom protocols; they are designed with partial programmability, known as DDP (Dynamic Device Personalization), and can be loaded with the customized firmware (not the default firmware) to support the additional telecom protocols.

NIC (1 queue) with CPU (1 core)

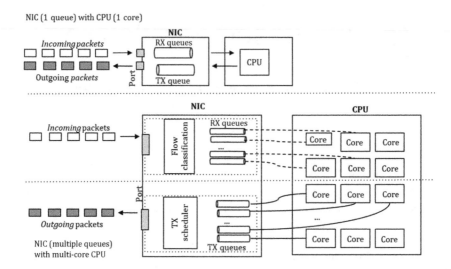

FIGURE 8.1 Multi-queue NIC and multi-core CPU.

Overall, NIC function is often limited by the silicon design phase. For the high-performance scenario, FPGA is a more flexible, customizable choice to support the advanced flow classification and is an important recipe to build the Smart NIC. FPGA can deliver high performance, its cost is relatively high, and it consumes more power.

8.1 MULTI-QUEUE

8.1.1 NIC Support

As we repeatedly say, in the silicon evolution in the past 15 years, the NIC port speed has evolved through 100 Mbps, 1 Gbps, 10 Gbps, 25 Gbps, 40 Gbps, and 100 Gbps. For the CPU, the number of cores increased dramatically, but CPU frequency stayed almost the same. Now, a single core is under challenge to serve a traffic of 10 Gbps and above, such as packet I/O and the additional application processing. One obvious choice is to take advantage of more cores available on the CPU, so that higher packet processing can be achieved, and this drives the NIC to come up with multi-queue—an efficient way to use the precious CPU core and cycles.

Intel® introduced the multi-queue feature on NIC 82575 (1 GbE) and later 82598 (10 GbE) in 2007. As the name suggests, the NIC RX/TX path has multiple hardware queues, and each queue can be associated with a specific core. In order to use the multiple queues, NIC came up with a few mechanisms (which will be discussed later) to distribute the incoming packets to the queues, and the actual packet move to the host memory is done with NIC DMA (direct memory access). When the CPU handles the packet, the packet and its buffer descriptor need to be loaded into cache. This memory and cache procedure is further optimized by the Intel® DDIO technology, and it was released together with the Sandy Bridge server platform in 2012.

Today, all Intel® NICs and the Linux kernel NIC drivers support the multi-queue feature, the number of supported queues is different in the different NICs, and the trend is to add more queues to meet the cloud-based use model. The network service can run in the virtual machine, which means the high-throughput traffic needs to come into the guest. For the guest workload, multiple cores can be assigned to process more packets to meet service demand. From the guest workload perspective, it is similar to bare-metal workload, and the multi-queue mechanism is very useful. The multi-queue can be assigned to the guest, supported by the physical NIC. In addition to physical NIC, the virtual NIC is also a popular interface, and virtio multi-queue is such an example, which will be discussed later (Table 8.1).

In addition to the load balancing use case, the multi-queue feature can be associated with target application processing, and certain packets can be directed into the specific queue(s), so that a specific protocol service can be used.

Multi-queue needs to be integrated with software enablement, and for user space drivers such as DPDK, the Ethernet PMD interface is built with the multi-queue support, and the number of queues needs to be decided when initializing/configuring the NIC port. The number of queues can be different at the RX/TX sides. The system designer will decide the following:

- How many queues need be configured?
- How many cores can be used for NIC driver (RX/TX)?
- How many cores are available for other packet processing software use?
- What are the packet distribution methods to send packets from the I/O core to different worker cores (for load balance processing, or application steer processing)?

The NIC can do the load balancing and direct the incoming traffic into the multiple queues, so that multiple cores can serve the queues in parallel. This scalable approach is popular and widely used on an x86 system.

Beyond the load balance, application-specific traffic steering is another use case. Take IPsec as the application example, where there is a need to handle only the IPsec packet and there is no need to process the other packet by the IPsec application. The NIC can assign one queue or multiple queues to IPsec only. IPsec packets are recognized by the NIC and moved to the specific queue, and then, the CPU can run the IPsec software. DPDK "rte_flow" API is the known software interface to configure NIC, and it does a similar job.

TABLE 8.1
Intel® NIC: Port, Driver, and Queues

Intel® NIC	Port Speed	SW Driver	NIC/Port Queues (RXQ/TXQ)	Max RXQ/TXQ to a VM
500 Series (82599/x550)	10 GbE/1 GbE	ixgbe	128/64	8/8, up to 64 VMs, each with two queues
700 Series (X710/XL710/ XXV710)	10/25/40 GbE	i40e	1024/1024	16/16, up to 128 VMs

How does the NIC distribute the incoming packets to multiple queues exactly? Most Ethernet devices have similar features, but the implementation is indeed vendor-specific. From the same vendor, the different generations of NIC products may have different features, and it is highly recommended to look into the NIC datasheet. Take the Intel® 10-GbE NIC (82599) as an example. Each port supports 64 virtual machines, and this NIC has 128 RX queues and 128 TX queues. A queue pair is a pair of a RX queue and a TX queue; hence, this NIC has 128 queue pairs.

The steps involved in the packet-receiving process are summarized below:

1. Recognizing a packet arrived on the Ethernet port
2. Filtering the packet on L2 Ethernet protocols
3. Queue assignment (such as Pool Select and Queue Select)
4. Moving the packet into the receive data FIFO
5. Transferring the packet into the host memory, owned by the specified queue
6. Updating the status of the packet buffer descriptor (Figure 8.2).

Packet filtering can happen to assess whether the packet is "good", and this means the packet is validated; for example, it comes with an allowed packet size and no CRC error. Then, the packet filter decision can be based on the L2 header: MAC address, VLAN, Ethernet type, and the combination. Decisions about whether to permit or deny a packet are subject to software configuration. If a packet is permitted into the NIC, the next step is "RX queue assignment", a new decision on assigning a packet into a queue (Pool Select, Queue Select).

Distributing the packets to queues is a critical step. Pool selection is Intel® 82599-specific hardware terminology, it is associated with virtualization use case, NIC can be configured in a way that resource is partitioned into PFs/VFs (in SR-IOV term), VFs can be assigned into a specific guest (tenant), pools can be mapped to

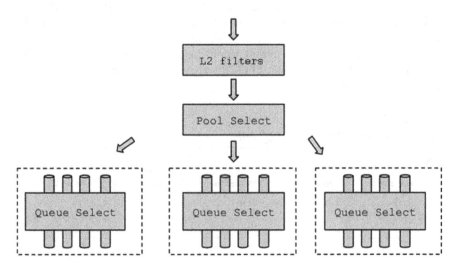

FIGURE 8.2 Intel® 82599: multi-queue selection process.

VFs, and each pool can have multiple queues. This "Pool Select" extends the NIC's multiple queues into a host/guest provision. In fact, the guest can have multiple cores, which work with multiple queues; this is useful for "high performance network service is running within a tenant", and this is useful for cloud-based network service.

For queue selection (queue assignment), RSS (Receive-Side Scaling) and Flow Director (FDIR) are important concepts. RSS is designed to distribute packets to multiple queues. A NIC uses a hash algorithm to compute a hash value over a defined received packet header (hash type). FDIR, initially introduced by Intel®, distributes packets to specific queues according to a match on certain packet header fields, and this is important for application steering. In addition to this, the NIC can also provide support for QoS according to priority-based queue assignment.

8.1.2 Linux Support

For a NIC with a single queue, the software has to support the traffic load balance to multiple cores. The Linux kernel handles this with the Receive Packet Steering (RPS) mechanism. RPS is mainly responsible for distributing software interrupts across CPU cores. The NIC driver will generate a hash identifier for each packet flow. The hash value can be derived from the packet header such as source/destination IP address, Layer 4 source/destination port, the derived hash value modulus the number of cores, which determines the core chosen as the interrupt target for the packet. In this way, the network traffic can be served with multiple cores. On the TX side, the packets, regardless of the CPU they come from, can be sent to the single TX queue supported by the NIC.

Later, this software load balancing is implemented on NIC, known as Receive-Side Scaling, and the NIC can parse the packet header to send them to the NIC multi-queue.

Let's take the Linux kernel as an example to see how NIC multi-queue is utilized. Here, we ignore the higher application layer processing; instead, we focus on the lower layer such as packet forwarding, which is implemented within the kernel, in which Linux NAPI and qdisc are widely applied. The data structure is multi-queue-ready.

A NIC device is represented by a struct net_device. It will contain multiple RX/TX queues. A network packet is stored in struct sk _ buff (known as "skb").

8.1.2.1 NIC RX

The NIC driver sets an interrupt request (IRQ) identifier for each RX queue. When a packet arrives, an IRQ is delivered to a core, which will serve the interrupt with software routine; essentially, it decides which core to handle the incoming packet. In early Linux systems, interrupts can be sent to a specific core (such as core 0). If all (or even most) interrupts are sent to core 0, it will be congested, and other cores may stay idle. So, it is not a balanced way. When lots of packets come to the multiple queues of NIC, a better way is to send IRQs to the multiple cores, and then, those cores can serve the interrupts and serve the multiple queues together at the RX side. IRQ affinity (smp_affinity) is such a mechanism, each IRQ can be allocated its affinity and assigned to a set of cores (using bit mask), and these affinitized cores can take care of the interrupts. Take the Linux example below, where the test environment has 4 cores (CPU0~CPU3). to The below command helps to find which interrupt and cores are involved here.

Query Interrupt

```
# cat /proc/interrupts
     CPU0      CPU1   CPU2    CPU3
128: 161392   135    188     62 IR-PCI-MSI-edge iwlwifi
```

IRQ #128 is dedicated to the Wi-fi network device, and all 4 cores serve interrupts.

Query the IRQ affinity of interrupt #128:

```
#cat /proc/irq/128/smp_affinity
f
```

Set affinity: Connect IRQ #128 to CPU3:

```
#echo 8 > /proc/irq/128/smp_affinity
#cat /proc/irq/128/smp_affinity
8
```

The Linux utility "ethtool" can query and configure the NIC device for the multi-queue, and channel is a similar concept to queue. The utility is also used for virtual NIC, such as virtio multi-queue, which will be discussed later.

```
-l --show-channels: It queries the specified network device
for the numbers of channels it has. A channel is an IRQ and
the set of queues that can trigger that IRQ.
-L --set-channels: It changes the numbers of channels of the
specified network device.
```

8.1.2.2 NIC TX

Linux provides a flexible queue selection mechanism. The dev _ pick _ tx function is used to select the TX queue. The driver can select a queue, or it can send packets with multiple TX queues (such as computing a hash of the sk _ buff). The XPS (Transmit Packet Steering, introduced after Linux kernel 2.6.38) mechanism can intelligently select one of the multiple queues to send the packet. A mapping from the core to a TX queue is available. By doing so, lock contention will be reduced because it reduces the competition for the same queue. The lock contention will be totally eliminated if each core has its own TX queue. DPDK has adopted this preferable approach as well. Second, the chance of cache miss is decreased during packet transmission. The following kernel code plainly illustrates that the selection of queues is taken into account during packet transmit:

```
int dev_queue_xmit(struct sk_buff *skb)
```

```
{
        struct net_device *dev = skb->dev;
        txq = dev_pick_tx(dev, skb);  //Select a queue.
        spin_lock_prefetch(&txq->lock);
        dev_put(dev);
}

struct netdev_queue *netdev_pick_tx(struct net_device *dev,
                                    struct sk_buff *skb)
{
        int queue_index = 0;

        if (dev->real_num_tx_queues != 1) {
          const struct net_device_ops *ops = dev->netdev_ops;
          if (ops->ndo_select_queue)
            queue_index = ops->ndo_select_queue(dev, skb);
          else
            queue_index = __netdev_pick_tx(dev, skb);
            queue_index = dev_cap_txqueue(dev, queue_index);
        }

        skb_set_queue_mapping(skb, queue_index);
        return netdev_get_tx_queue(dev, queue_index);
}
```

The TX and RX queues are generally assigned to the same interrupt vector, so the same core may handle both RX/TX interrupts; this helps the cache utilization if a packet is received from RX queue 1 is sent out on TX queue 1, by a same core.

Linux also provides a range of interfaces to use multi-queue NICs. For an Intel® CPU, each core has its local L1 and L2 cache, not shared, and all cores share a common L3 cache, which has a much bigger size. The effective use of the cache improves the system performance. The software designer should pay attention to ensure the packet data move is very effective.

8.1.3 DPDK SUPPORT

Linux has a long evolution path, it supports CPU evolved from single core to multi-core, and it supports the NIC from single queue to multi-queue. DPDK was born later, is much simpler, is always assumed to run in a multi-queue, multi-core hardware platform, and uses the poll node for NIC, and the dedicated cores are assigned (by system designer) to handle the packets that arrive on NIC and are responsible for sending the packets to the multiple queues at the TX side.

The packet I/O mechanism is inherently designed with multi-queue and multi-core. DPDK APIs are written with high parallelism in mind. By default, DPDK will initialize the NIC device with the multi-queue, it will configure the queue and bind it with CPU cores, and it is designed to run in a NUMA-aware platform. The l3fwd sample provides a good reference in the early chapters. DPDK principles are listed as follows:

- Each NIC port will be configured with multi-queue at both the RX and TX sides, owned by DPDK PMD.
- DPDK supports the port, queue, and core binding mechanism.
- If a RX queue of the NIC is assigned to a core, the packets received from the queue are assumed to be processed completely on that core. This is the run-to-completion model.
- If the memory pool is allocated from the local memory channel, closer to the core, both the received packet and the buffer descriptor should be located in the same memory pool.
- If a TX queue is assigned to the core, both the packet and its buffer descriptor should be located in the local memory pool corresponding to the core and the TX queue.

If each queue is accessed by one dedicated core owner, it avoids the excessive lock overhead from multi-core access. A code snippet taken from the l3fwd sample which describes the port, queue, and core binding approach is given below:

```
nb_tx_queue = nb_lcores;
      ......

ret = rte_eth_dev_configure(portid, nb _ rx _ queue,
            (uint16_t)nb _ tx _ queue, &port_conf);

queueid = 0;
for (lcore_id = 0; lcore_id < RTE_MAX_LCORE; lcore_id++) {
      if (rte_lcore_is_enabled(lcore_id) == 0)
      continue;
......
ret = rte _ eth _ tx _ queue _ setup(portid, queueid, nb_txd,
socketid, txconf);
......

qconf->tx_queue_id[portid] = queueid;
      queueid++;
      }

/* In the thread corresponding with the logical core: */
queueid = qconf->tx_queue_id[port];
......

ret = rte_eth_tx_burst(port, queueid, m_table, n);
```

In the real world, CPU cores are expensive resources, while NIC has many queues and is less expensive, so it is common to assign one core to serve many queues. It is rare that the number of cores is greater than the number of TX queues, more than one core can be assigned to the queue, and the use of lock is required for data synchronization.

8.2   FLOW CLASSIFICATION

Flow classification is how a NIC receives, parses, and classifies the arrived packet
header with an action. The classification result can be written into the received packet
descriptor. Because this is done by the NIC, it saves CPU cycles to do a similar job with
software. Flow classification rules and actions are often vendor-specific (Table 8.2).

8.2.1   PACKET TYPE

NIC can read the incoming packet header to recognize "many" packet types and
write the packet type into the buffer descriptor. This can save software-based parsing cycles. Intel® XL710, for example, can recognize many packet types such as IP,
TCP, UDP, VXLAN, and NVGRE, and the software can use this feature. There are
so many networking protocols, so not all packet types can be recognized by the
NIC.

In DPDK, each packet buffer descriptor includes the fields given below, which are
in the data structure of "rte _ mbuf". The below code shows packet _ type at
different layers such as L2, L3, and L4 and tunnel packet information.

```
struct rte_mbuf {
        ......
        union {
                uint32_t packet_type; /**< L2/L3/L4 and tunnel
                information. */
                struct {
                        uint32_t l2_type:4; /**< (Outer) L2 type. */
                        uint32_t l3_type:4; /**< (Outer) L3 type. */
                        uint32_t l4_type:4; /**< (Outer) L4 type. */
                        uint32_t tun_type:4; /**< Tunnel type. */
                        uint32_t inner_l2_type:4; /**< Inner L2
                        type. */
                        uint32_t inner_l3_type:4; /**< Inner L3
                        type. */
```

TABLE 8.2
Flow Classification, Match, and Actions

Technique	Matching Information	Action
Packet type	Protocol info	Mark packet type
SR-IOV/VMDQ	MAC address, VLAN ID	Mark to queue set
RSS	IP address, port ID...	Director to queue
Flow Director	IP address, port ID...	Director to queue/drop
QoS	UP in VLAN tag (3-bits)	Mark to queue set
Legitimacy verification	MAC address, VLAN ID	Mark to queue set/director to queue/director to control engine/drop

```
                uint32_t inner_l4_type:4; /**< Inner L4
                type. */
        };
    };
    ......
}
```

8.2.2 Receive-Side Scaling (RSS)

The RSS concept is discussed earlier. As a NIC hardware feature designed for load balance use case, once enabled, RSS can distribute packets to different queues, and it is based on a hash calculation on certain packet header fields. Intel® XL710 NIC is given as an example, and on a given NIC port, RSS can be configured as follows (Figure 8.3):

- First, the packet type should be selected; then, depending on the different packet types, the different packet input fields can be used for hash computation.
- Hash computation is customizable, though the algorithm is implemented on NIC.
- Hash value determines the queue index, which directs the packet into a specific queue. It is determined by a table that maps from the hash value calculated from the packet's fields to the queues in the hardware.

It is hard to predict whether RSS will evenly distribute packets to queues. It depends on the incoming packets, and it also depends on the hash algorithm on a NIC. Intel® XL710 supports multiple hash functions. Generally, Microsoft's Toeplitz hash is a popular hash function. In some network systems, a symmetric hash algorithm is preferred for performance reasons. A symmetric hash is to

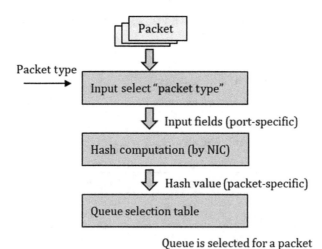

FIGURE 8.3 RSS flow on a NIC.

ensure hash (source address, destination address) == hash (destination address, source address). This is very useful for the bidirectional packet flow, such as TCP connection; the bidirectional flow is indeed the same data flow, and it is better to be processed on the same core. All other stateful connections may prefer a similar model. If each direction of the connection is processed by different cores, there will be data synchronization among different cores, resulting in additional overhead.

The packet type decides what RSS inputs are used for hash computation input and is NIC-specific, and the supported input fields for Intel® XL710 are listed in the table below. For example, RSS input fields for the IPv4 UDP packet are as follows:

- Source IP address (S-IP),
- Destination IP address (D-IP),
- Source port number (S-Port),
- Destination port number (D-Port) (Table 8.3).

RSS input fields for an IPv4 packet with an unknown protocol are as follows:

- Source IP address (S-IP),
- Destination IP address (D-IP).

DPDK defines the basic RSS configuration interface for NIC, the RSS Key (and its length and hash functions are vendor-specific; always refer to the datasheet). Software tuning may lead to the even packet distribution among RX queues, which is only required if the default configuration is not working properly.

```
struct rte_eth_rss_conf {
    uint8_t *rss_key; /**< If not NULL, 40-byte hash key. */
    uint8_t rss_key_len; /**< hash key length in bytes. */
```

TABLE 8.3
RSS Input Fields for Intel® XL710

Packet Type	Hash Field Input
IPV4 UDP	S-IP, D-IP, S-Port, D-Port
IPV4 TCP	S-IP, D-IP, S-Port, D-Port
IPV4 SCTP	S-IP, D-IP, S-Port, D-Port, Verification-Tag
IPV4 OTHER	S-IP, D-IP
IPV6 UDP	S-IP, D-IP, S-Port, D-Port
IPV6 TCP	S-IP, D-IP, S-Port, D-Port
IPV6 SCTP	S-IP, D-IP, S-Port, D-Port, Verification-Tag
IPV6 OTHER	S-IP, D-IP

After the packets are placed into multiple queues, the multiple cores will process the packets, which mostly run the same processing.

8.2.3 FLOW DIRECTOR (FDIR)

Flow Director originated from Intel®, and the NIC can distribute packets to different queues according to a match on certain packet header fields (such as n-tuples). For RSS, most assigned cores (serving the RX queues) are doing similar processing. For FDIR use case, the assigned cores may work on different processing, because the RX queue may get different packet flow; for example, IPsec flows can be sent into some queue, and UDP traffic can be directed into other queues. This approach can enable the application target processing, and the target application may depend on the destination port or IP address. Figure 8.4 shows how it works.

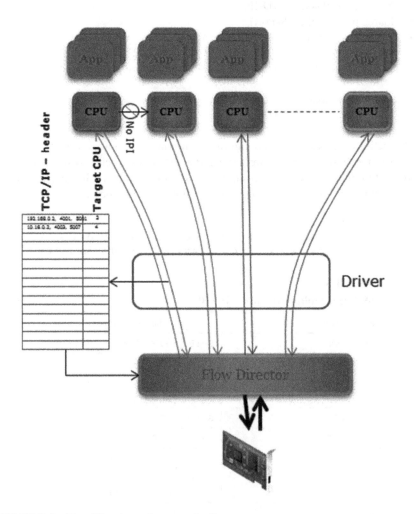

FIGURE 8.4   Flow Director and queue selection.

A Flow Director table is built on a NIC, its size is limited by hardware, the table entry provides the "input fields" to be matched against the incoming packets, and the action after matching is also defined in the table entry, such as which queue to receive the arrived packet. The Linux software utility ethtool can be used for the Flow Director setup, and the steps for Intel® Ethernet adapters are as follows:

```
# ethtool --show-features <interface name> | grep ntuple
/*Is fdir supported? */
# ethtool --features <interface name> ntuple on
/* Turn on Flow director */

# ethtool --config-ntuple flow-type tcp4 src-ip 10.23.4.6
dst-ip 10.23.4.18 action 4
/* This traffic from 10.23.4.6 to 10.23.4.18 be placed in
queue 4 */
```

The NIC parses the incoming packet header to get the packet type and will look up the Flow Director table; if "input fields" match a flow table entry, the defined action will be taken. Similar to RSS, the actual input fields are subject to the packet type. The below table lists the mapping between input fields and packet types on Intel® XL710. The user can also specify the optional fields on a given packet type. For example, if the incoming packet is recognized as IPv4 UDP, the flow matching can just look up the destination port (D-Port) only, and the other fields (S-IP, D-IP, and S-PORT) can be ignored (Table 8.4).

Flow Director enables the selected packet flow to specific queue, and it is a high granularity of packet control to serve the target application. In the network virtualization use case, the user may reserve a queue for a specific VXLAN-GPE session, and VNI (VXLAN Network Identifier) can be set as a flow. This approach can enforce the virtualized network traffic to arrive on a particular queue, and this queue can be assigned to the guest workload (regardless of whether it is a virtualization- or container-based cloud infrastructure).

TABLE 8.4
Flow Director: Input Fields for Intel® XL710

Packet Type	Fields
IPV4 UDP	S-IP, D-IP, S-Port, D-Port
IPV4 TCP	S-IP, D-IP, S-Port, D-Port
IPV4 SCTP	S-IP, D-IP, S-Port, D-Port, Verification-Tag
IPV4 OTHER	S-IP, D-IP
IPV6 UDP	S-IP, D-IP, S-Port, D-Port
IPV6 TCP	S-IP, D-IP, S-Port, D-Port
IPV6 SCTP	S-IP, D-IP, S-Port, D-Port, Verification-Tag
IPV6 OTHER	S-IP, D-IP

8.2.4 NIC SWITCHING

Only legitimate packets are received into the NIC. For the incoming packets, the "good" packets will be received to the host or guest memory. Switching capability is a feature for the cloud-based use case. As shown in the below figure, a system is running with the host and many virtual machines. The NIC has to be partitioned so that some VFs are assigned by the VMs.

For the incoming packets from the external system, the NIC will decide which VM will receive the packet. From a VM perspective, the outgoing packets may go out of NIC to another external system, or it may go to a VM running in the same host (such as VM0 <-> VM1). The packet forward decision can happen on software layer (usually done by virtual switch), or it can be off-loaded by a NIC (Figure 8.5).

If the NIC is embedded with an internal switch, sometimes it is called SR-IOV-based switching, and the packet forward decision can be made on a NIC; it compares the Ethernet header such as MAC/VLAN fields to do an internal switching, for example, from VM1 to VM2, because VM1 and VM2 can be placed in the same network. This is a Layer 2-based networking solution to interconnect VMs. In today's large-scale public cloud system, the network overlay is the new major approach, the packet forward

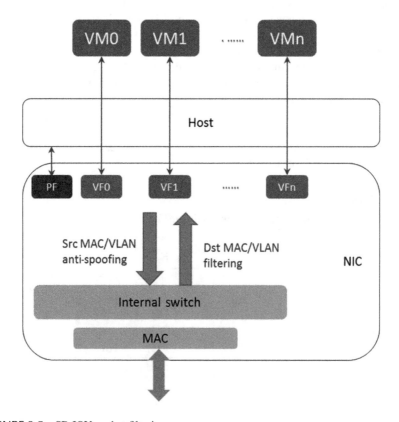

FIGURE 8.5 SR-IOV packet filtering.

decision among VMs is moved to Layer 3, and the IP address is used for routing the traffic to appropriate VM. As a result, the software-based virtual switching has been dominated in use by the production deployment. In recent years, Smart NIC is the latest trend, where the virtualized network can be fully off-loaded in a NIC. If the guest has the high I/O throughput requirement, the multi-queue is always a preferred approach.

8.2.5 Customized Flow Classification

If the I/O network device (NIC or FPGA) supports the customization, the system designer can add the additional capability for flow classification. Due to the historical reason that the NIC is usually designed for Ethernet/IP-based network stack, it will not support the in-depth packet header parsing and matching logic. Often NIC is not able to support the wireless data plane stack such as GTP-U. Some advanced NICs (such as Intel® XL710) are able to support the additional custom firmware; essentially, it configures the NIC with the new capability, and this additional feature is known as DDP (Dynamic Device Personalization), which helps the workload-specific optimization (Figure 8.6).

Without DDP, Intel® XL710 is not able to parse certain packet types; for example, PPPoE and GTP-U are not supported, which implies that NIC is limited to directing the PPPoE/GTP-U traffic into multiple queues. PPPoE is heavily used for the broadband access gateway, and GTP-U is heavily used for the wireless core gateway. DDP leverages the built-in packet pipeline parser in NIC, but it goes a little further on the packet header parsing. With DDP support, the additional firmware can be loaded into the NIC, and it adds the new capability to NIC at runtime. FPGA is a popular choice to make a Smart NIC, and using the FPGA programmability to add the advanced flow classification is also a common choice.

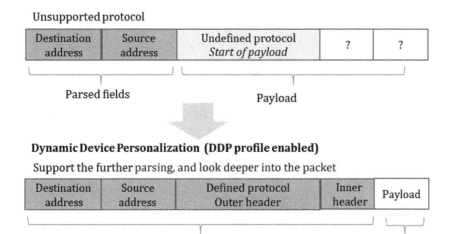

FIGURE 8.6 Intel® XL710: DDP.

8.3 USE CASE

Multi-queue NICs need to be combined with different flow classification techniques in the real world. Here, two use cases are illustrated below:

  Case 1: Hybrid use of RSS/FD,
  Case 2: Virtualized networking use case.

8.3.1 HYBRID RSS AND FDIR

Imagine a server system that needs to implement two planes: forwarding plane and control plane. The forwarding plane is used for large amounts of packet processing and requires simple processing but often high packet throughput, it is good to assign multiple cores, and RSS is preferred, which helps the high-performance scaling design.

The control plane is used for flow/session management and responsible for flow and session management. Usually, this is a small traffic but has high processing complexity, though not lots of computation cycles. Flow Director can be helpful to enable the application steering, and maybe one core is sufficient (Figure 8.7).

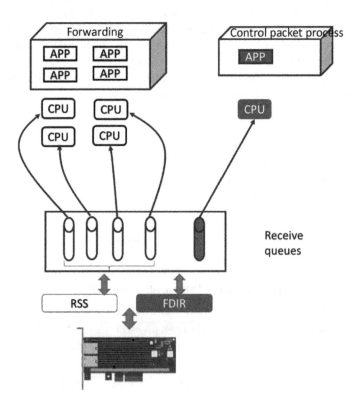

FIGURE 8.7 Hybrid RSS and FD use.

In this case, the basic ideas are as follows:

1. Divide the workload (forwarding plane, control plane) with multi-queue, multi-cores
2. Save core cycles by making a good use of NIC capabilities (RSS, FD)
3. Minimize the packet data movement among the application's cores.

In the below code snippet:

- Four RX queues are used on the forwarding plane: Queues 0, 1, 2, and 3 are served by four cores, for the forwarding plane
- Queue 4 is configured with Flow Director for one core on the control plane processing. Let's assume it targets with a UDP packet flow:

```
Source IP = 2.2.2.3,
Destination IP = 2.2.2.5,
Source Port = Destination Port = 1024,
This is shown in the gray queue.
```

1. Initialize the NIC:

RSS and Flow Director are configurable hardware resources on the NIC. How to enable RSS and Flow Director in DPDK is shown below:

```
static struct rte_eth_conf port_conf = {
  .rxmode = {
        .mq_mode = ETH_MQ_RX_RSS,
  },
  .rx_adv_conf = {
        .rss_conf = {
                .rss_key = NULL,
                .rss_hf = ETH_RSS_IP | ETH_RSS_UDP
                      ETH_RSS_TCP | ETH_RSS_SCTP,
        },
  }, //Enable RSS
  fdir_conf = {;
    .mode = RTE_FDIR_MODE_PERFECT,
  .pballoc = RTE_FDIR_PBALLOC_64K,
  .status = RTE_FDIR_REPORT_STATUS,
  .mask = {
    .VLAN_tci_mask = 0x0,
    .ipv4_mask    = {
    .src_ip = 0xFFFFFFFF,
    .dst_ip = 0xFFFFFFFF,
    },
    .ipv6_mask    = {
        .src_ip = {0xFFFFFFFF, 0xFFFFFFFF, 0xFFFFFFFF,
0xFFFFFFFF},
        .dst_ip = {0xFFFFFFFF, 0xFFFFFFFF, 0xFFFFFFFF,
0xFFFFFFFF},
      },
```

```
        .src _ port _ mask = 0xFFFF,
        .dst _ port _ mask = 0xFFFF,
        .mac _ addr _ byte _ mask = 0xFF,
        .tunnel _ type _ mask = 1,
        .tunnel _ id _ mask = 0xFFFFFFFF,
    },
    .drop _ queue = 127,
},//Enable the Flow Director.
};

rte _ eth _ dev _ configure(port,    rxRings,    txRings,
&port _ conf);
//Configure the device.
//In this case, rxRings = 5; txRings = 5.
```

2. Configure the TX and RX queues:

```
mbuf _ pool = rte _ pktmbuf _ pool _ create();
//Reserve the main memory for the packets RX/TX

for (q = 0; q < rxRings; q ++) {
  retval=rte _ eth _ rx _ queue _ setup(port, q, rxRingSize,
                            rte _ eth _ dev _ socket _
                            id(port),
                            NULL,
                            mbuf _ pool);
  if (retval < 0)
       return retval;
}

for (q = 0; q < txRings; q ++) {
  retval=rte _ eth _ tx _ queue _ setup(port, q, txRingSize,
                            rte _ eth _ dev _ socket _
                            id(port),
                            NULL);
  if (retval < 0)
       return retval;
}
//Configure the send and receive queues.
//In this case, rxRings = 5; txRings = 5.
```

3. Start the Ethernet device

```
rte _ eth _ dev _ start(port);
```

4. Set up the Flow Director

```
struct rte _ eth _ fdir _ filter arg =
{
  .soft _ id = 1,
```

```
.input = {
.flow _ type = RTE _ ETH _ FLOW _ NONFRAG _ IPV4 _ UDP,
.flow = {
  .udp4 _ flow = {
  .ip = {.src _ ip = 0x03020202, .dst _ ip = 0x05020202,}
            .src _ port = rte _ cpu _ to _ be _ 16(1024),
          .dst _ port = rte _ cpu _ to _ be _ 16(1024),
                  }
      // Prepare the Flow Director config
      }
  }
  .action = {
    .rx _ queue = 4,
    .behavior= RTE _ ETH _ FDIR _ ACCEPT,
    .report _ status = RTE _ ETH _ FDIR _ REPORT _ ID,
  }
}


r t e _ e t h _ d e v _ f i l t e r _ c t r l ( p o r t ,
RTE _ ETH _ FILTER _ FDIR,
RTE _ ETH _ FILTER _ ADD, &arg);
// Add the Flow Director entry to the table.
```

Reconfigure RSS:

The RSS configuration distributes packets evenly based on the number of receive queues. Queue 3 receives only a specific UDP data flow, and this is done by the Flow Director configured in the previous step, which will assign the specified UDP packets into Queue 3. RSS needs to be reconfigured; otherwise, RSS will balance the remaining non-specific UDP traffic to all queues (including Queue 3). The below step will remove Queue 3 out of RSS table.

```
// Configure the hash value queue number mapping table.
// The table size is 128 for 82599 NIC.
struct rte _ eth _ rss _ reta _ entry64 reta _ conf[2];
int i, j = 0;
for (idx = 0; idx < 3; idx++) {
  reta _ conf[idx].mask = ~0ULL;
  for (i = 0; i < RTE _ RETA _ GROUP _ SIZE; i++, j++) {
        if (j == 4)
              j = 0;
        reta _ conf[idx].reta[i] = j;
  }
}
rte _ eth _ dev _ rss _ reta _ query(port, reta _ conf, 128);
```

5. Then, each application thread can receive/send packets from/to their distributed queues.

8.3.2 Virtualization

Intel® XL710 NIC can be configured as follows:

- PF is managed by the Linux kernel driver (i40e)
- VFs, passed through to VMs, can be managed by DPDK (i40evf) driver.

The Linux "ethtool" utility can configure the Flow Director on the NIC, and the specific packets can be directed and sent to a given VF's queues. The packets will later be processed by the virtual network function (VNF) running in a VM (Figure 8.8).

NIC queues can be shared by the Linux kernel and DPDK driver, and they can be used by both the host and guests with the proper configuration. All of them can work together and in parallel. The below solution diagram is an example of PF and VF that the Linux kernel and DPDK driver resource share on a NIC. This solution is valuable because of the following reasons:

1. There is full Linux support with network device management with PF
2. There is full Linux network protocol support on the packets that are routed to Linux kernels via queues
3. VF is directly assigned to the guest.
   - If VM needs to support the high network throughput, the driver can be optimized by DPDK VF PMD, and this can be a VNF use case. VNF can be launched with the cores
   - If VM does not require the high network throughput, the guest can skip using DPDK VF in the guest; instead, the guest can just use the standard Linux networking stack.

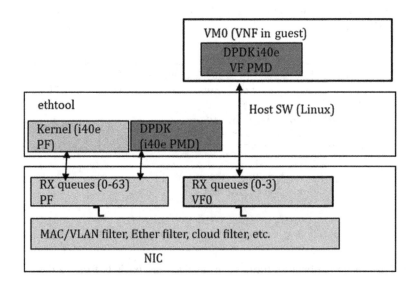

FIGURE 8.8   Host- and virtualization-based multi-queue uses.

A few steps to implement it are as follows:

1. In system BIOS, enable Intel® VT;
2. Add Intel_iommu=on to the Linux kernel boot command line.
3. Install Linux kernel driver (i40e.ko for XL710) for customerization, and download source codes of the i40e-1.3.46 driver

```
CFLAGS _ EXTRA="-DI40E _ ADD _ CLOUD _ FILTER _ OFFLOAD"
make
rmmod i40e
insmod i40e.ko
```

4. Create VF0

```
echo 1 > /sys/bus/pci/devices/0000:02:00.0/sriov _ numvfs
modprobe pci-stub
echo "8086 154c" > /sys/bus/pci/drivers/pci-stub/new _ id
echo  0000:02:02.0  >  /sys/bus/pci/devices/0000:2:02.0/
driver/unbind
echo 0000:02:02.0 > /sys/bus/pci/drivers/pci-stub/bind
```

5. Start VM with VF0

```
qemu-system-x86 _ 64-name  vm0-enable-kvm  -cpu  host  -m
2048-smp  4-drive  file=dpdk-vm0.img  -vnc:4-device  pci-
assign, host=02:02.0
```

6. Configure NIC on the host (via ethtool), and NIC will direct the packet flow to VF:

```
ethtool  -N  ethx  flow-type  ether  dst  00:00:00:00:01:00  m
ff:ff:ff:ff:ff:ff  src  00:00:00:00:20:00  m  00:00:00:00:00:00
vlan 10 user-def 0x800000000 action 2 loc 1
```

It is expected to see the below packets that will be received on Queue 2 of VF:

- Inner destination MAC address = 00:00:00:00:20:00,
- Inner VLAN = 10,
- VNI = 8
  1. If the upper 32 bits of "user-def" is 0xffffffff, then the filter can be used for programming an L3 VEB filter; otherwise, the upper 32 bits of "user-def" can carry the tenant ID/VNI if specified
  2. Cloud filters can be defined with inner MAC, outer MAC, inner IP, inner VLAN, and VNI as part of the cloud tuple. It is always the destination (not source) MAC/IP that these filters use. For the above example, where both dst and src address fields are loaded, dst == outer and src == inner.
  3. The filter will direct a packet matching the rule to a VF ID specified in the lower 32 bits of "user-def" to the queue specified by "action".

4. If the VF ID specified by the lower 32 bits of "user-def" is greater than or equal to max_vfs, then the filter is for the PF queues.

Such a solution is also common for cloud data center infrastructure, and there is a general requirement to separate the data traffic for the computing traffic and storage traffic. Depending on the system workload, the computing traffic may go through DPDK-accelerated vSwitch to support the virtualized network. The storage traffic may just go to the Linux kernel-based software. For more details on SR-IOV, vSwitch, and VNF, refer to the later chapters.

8.3.3 Rte _ flow

Rte_flow is the DPDK API to configure the packet flow, it has been introduced since DPDK v17.02, and the intent is to provide a generic means for flow-based configuration, so that the hardware can handle the ingress or egress traffic at the flow level. NIC is very different, and a generalized software interface is desired to work with all PMDs. As a result, rte_flow API is at a somewhat higher level, and it is used to configure NIC or even FPGA.

The below example (test-pmd) demonstrates how to create flow rules for a NIC (Intel® XL710) with two approaches: Configure via command lines, or configure via rte_flow source call.

Let's start first with the command lines.

1. Bind NIC to igb_uio

```
./usertools/dpdk-devbind.py -b igb _ uio 0000:06:00.0
```

2. Create a VF on DPDK PF; then, bind it to vfio-pci

```
echo 1 > /sys/bus/pci/devices/0000:06:00.0/max _ vfs
./usertools/dpdk-devbind.py -b vfio-pci 0000:06:02.0
```

3. Launch two test-pmd instances with PF and VF separately

```
./x86 _ 64-native-linuxapp-gcc/app/test-pmd  -c 1ffff  -n
4--socket-mem 1024,1024--file-prefix=pf  -w  06:00.0--  -i
--rxq=16--txq=16
testpmd> rx _ vxlan _ port add 4789 0
testpmd> set fwd rxonly
testpmd> set verbose 1
testpmd> start

./x86 _ 64-native-linuxapp-gcc/app/testpmd   -c  1e0000-n
4--socket-mem 1024,1024--file-prefix=vf  -w  06:02.0--  -i
--rxq=4--txq=4
testpmd> set fwd rxonly
testpmd> set verbose 1
testpmd> start
```

4. Create the flow rule—direct VXLAN (tunnel) packets to VF0/Queue 3

```
testpmd> flow create 0 ingress pattern eth dst is 00:11:22:
33:44:66 / ipv4 / udp / vxlan vni is 7 / eth dst is 00:11:22:33:
44:55 / end actions vf id 0 / queue index 3 / end
```

5. Send the VXLAN packet via Scapy (Scapy is a test utility)

```
sendp(Ether(dst="00:11:22:33:44:66")/IP()/UDP()/
VXLAN(vni=7)/Ether(dst="00:11:22:33:44:55")/IP()/TCP()/
Raw('x' * 20), count=1, iface="ens802f0")
```

6. Validate if the VXLAN packet is received on VF0/Queue 3

```
test-pmd> port 0/queue 3: received 1 packet
src=00:00:00:00:00:00- dst=00:11:22:33:44:66- type=0x0800-
length=124- nb _ segs=1- hwptype: L2 _ ETHER L3 _ IPV4 _ EXT _
UNKNOWN TUNNEL _ GRENAT INNER _ L2 _ ETHER INNER _ L3 _
IPV4 _ EXT _ UNKNOWN INNER _ L4 _ TCP - sw ptype: L2 _ ETHER
L3 _ IPV4 L4 _ UDP
- 12 _ len=14- 13 _ len=20- 14 _ len=8- VXLAN packet: packet
type =24721, Destination UDP port =4789, VNI = 7- Receive
queue=0x3
ol _ flags: PKT _ RX _ L4 _ CKSUM _ GOOD PKT _ RX _ IP _ CKSUM _
GOOD PKT _ RX _ OUTER _ L4 _ CKSUM _ UNKNOWN
```

The above flow creation (in Step 4) is equivalent to the following software code using rte_flow APIs:

```
struct rte_flow_attr attr;
struct rte_flow_item pattern[MAX_PATTERN_NUM];
struct rte_flow_action action[MAX_ACTION_NUM];
struct rte_flow *flow = NULL;
struct rte_flow_action_vf vf = { .id = 0 };
struct rte_flow_action_queue queue = { .index = 3 };
struct rte_flow_item_vxlan vxlan_spec = { .vni = "\x7" };
struct rte_flow_item_eth outer_eth_spec = {
    .dst.addr_bytes = "\x00\x11\x22\x33\x44\x66",
    };

struct rte_flow_item_eth inner_eth_spec = {
    .dst.addr_bytes = "\x00\x11\x22\x33\x44\x55",
    };

struct rte_flow_item_eth eth_mask = {
    .dst.addr_bytes = "\xff\xff\xff\xff\xff\xff",
    };

int res;
```

```
memset(pattern, 0, sizeof(pattern));
memset(action, 0, sizeof(action));

/* set the rule attribute.
 * in this case only ingress packets will be checked.
 */
memset(&attr, 0, sizeof(struct rte_flow_attr));
attr.ingress = 1;

/* set the match patterns, eth/ipv4/udp/vxlan/eth.*/
pattern[0].type = RTE_FLOW_ITEM_TYPE_ETH;
pattern[0].spec = &outer_eth_spec;
pattern[0].mask = &eth_mask;

pattern[1].type = RTE_FLOW_ITEM_TYPE_IPV4;
pattern[2].type = RTE_FLOW_ITEM_TYPE_UDP;

pattern[3].type = RTE_FLOW_ITEM_TYPE_VXLAN;
pattern[3].spec = &vxlan_spec;
pattern[3].mask = &rte_flow_item_vxlan_mask;

pattern[4].type = RTE_FLOW_ITEM_TYPE_ETH;
pattern[4].spec = &inner_eth_spec;
pattern[4].mask = &eth_mask;

/* the final level must be always type end */
pattern[5].type = RTE_FLOW_ITEM_TYPE_END;

/* create the action sequence.
 * move packet to specific queue of vf
 */
action[0].type = RTE_FLOW_ACTION_TYPE_VF;
action[0].conf = &vf;
action[1].type = RTE_FLOW_ACTION_TYPE_QUEUE;
action[1].conf = &queue;
action[2].type = RTE_FLOW_ACTION_TYPE_END;

/* validate and create the flow */
res = rte_flow_validate(port_id, &attr, pattern, action,
error);
if (!res)
   flow = rte_flow_create(port_id, &attr, pattern, action,
error);
```

8.4   CONCLUSION

NIC implementation is vendor-specific, and load balancing and flow classification are basic features. Making best use of NIC features can improve system performance, but it requires technical knowledge to have a full understanding of these features and maximize the benefit of NICs. Overall, higher speed and more intelligent flow processing are a trend for the future NICs. The cloud data center expects to deploy many applications based on network virtualization, which requires security filtering and differentiated service policies for the incoming and outgoing packets. Intelligent (Smart) NICs are expected to help realize more complicated network features with less computing resources.

9 NIC Offload

Wenzhuo Lu and Heqing Zhu
Intel®

CONTENTS

The previous chapter discussed NIC features such as flow/traffic classification with multi-queue, which can be used together with multi-cores for high-performance packet I/O.

In this chapter, NIC-based hardware off-loading will be further discussed. NIC offload can save CPU cycles; for example, it can help the VLAN stripping and insertion and packet checksum computation or even support the TCP segmentation.

9.1 NIC OFFLOAD

Ethernet and TCP/IP are the main protocols used by Internet networking. NIC can read and then parse the arrived packet header:

- Ethernet/IP packet integrity check,
- Protocol parsing in multiple layers (Ethernet/VLAN, IPv4/IPv6, TCP/UDP/SCTP)
- Support network overlay protocols such as VXLAN.

Each packet is different and has its own packet buffer descriptor, the NIC will write the header parsing result into the buffer descriptor, and it is subject to the hardware definition. The NIC driver is responsible for setting up the buffer descriptor and packet buffer. DPDK supports all major Ethernet silicon drivers, and the PMDs are available in open source, so all code details can be found in source code format.

Due to the limited knowledge of the author, only a subset of Intel® NICs are given as examples here to describe the NIC offload feature. The below table briefly lists the supported offload features of Intel® NICs: i350, 82599, x550, and xl710 (Table 9.1).

TABLE 9.1

Intel® NICs: Offload Features

Item	Function	i350	82599	X550	Xl710
Port speed	GbE	1	10	10	25/40
Packet header offload	VLAN Insert/Strip	√	√	√	√
	Double VLAN Insert/Strip	√	√	√	√
	IEEE1588	√	√	√	√
	Checksum offload for IP/TCP/UDP/SCTP	√	√	√	√
	VXLAN and NVGRE			√	
Segmentation (TSO)	TCP segmentation offload TX only	√	√	√	√
Coalescing (RSC)	Receive-side coalescing. RX only			√	
DDP	Additional protocol parsing support				√

9.2 OFFLOAD FLAGS

NIC offload features are often configurable on a port basis. It allows the different configuration for different ports. Only the packets going through the port are enforced by the port-based offload feature. In DPDK, every packet can be placed into a single mbuf, or multiple mbufs (for large packets). As said, each packet buffer (rte _ mbuf) has 64 bits for offload flags (ol _ flags). DPDK software reads or writes the offload flags for every packet, so this is important for each packet reception and transmission with NIC interactions. It includes the instructions to guide NIC on how to handle every packet (or even differently).

At the RX side, some supported bits in ol_flags (not a complete list) (Table 9.2).

At the TX side, some supported offload flags (not a complete list) (Table 9.3).

TABLE 9.2

NIC RX Offloads

ol_flags in Bits	Description
PKT_RX_VLAN_PKT	If bit is set, the RX packet has VLAN, and VLAN ID is stripped to rte_mbuf
PKT_RX_RSS_HASH	If bit is set, RSS hash result is available for RX packet in rte_mbuf
PKT_RX_FDIR	If bit is set, the Flow Director result is available in rte_mbuf
PKT_RX_L4_CKSUM_BAD PKT_RX_IP_CKSUM_BAD	Checksum validation result on the RX packet side.
PKT_RX_IEEE1588_PTP; PKT_RX_IEEE1588_TMST	IEEE1588 off-loading

TABLE 9.3
NIC TX Offloads

ol_flags in Bits	Description
PKT_TX_VLAN_PKT	If bit is set, insert VLAN ID in rte_mbuf before send packet goes out
PKT_TX_IP_CKSUM PKT_TX_TCP_CKSUM PKT_TX_SCTP_CKSUM PKT_TX_UDP_CKSUM PKT_TX_OUTER_IP_CKSUM PKT_TX_TCP_SEG PKT_TX_IPV4 PKT_TX_IPV6 PKT_TX_OUTER_IPV4 PKT_TX_OUTER_IPV6	If bit is set, checksum is computed and inserted into the packet header. These flags can be used in the TSO and VXLAN/NVGRE protocol scenarios
PKT_TX_IEEE1588_PTP;	IEEE1588 off-loading

Offload flags are under continuous development, and their details can be found at http://www.dpdk.org/browse/dpdk/tree/lib/librte_mbuf.

9.3 OFFLOAD CAPABILITY

There are a variety of NICs; hence, the supported offload features in different NICs also vary. Hardware features have to be enabled by software. DPDK focuses on the generalized offload feature list.

1. There might be cases that certain offload functions are available on new NIC, but not supported by DPDK yet. To remove the gap on the latest NIC, the additional DPDK development might be required.
2. DPDK APIs provide the software abstraction of different NICs. It does not take into account the fact that some NICs support features while others do not. Therefore, the suggestion is to query the offload capability first.

To use a NIC offload feature, the application will call the DPDK software interface, it will go through PMD to access the actual hardware, and the software interface can be at the port level or packet level. PMD provides the globally visible offload features based on the port level, and some offload features are configurable. Using "test-pmd" can query the offload features and the configuration.

```
test-pmd> show port <port_id> rx_offload capabilities
test-pmd> show port <port_id> rx_offload configuration
test-pmd> show port <port_id> tx_offload capabilities
test-pmd> show port <port_id> tx_offload configuration
```

Each individual packet has its own memory buffer (rte_mbuf), which includes the buffer descriptor and packet memory, and the buffer descriptor reserves the fields for packet-specific offload results at the RX side, and the packet instruction to guide the NIC at the TX side.

Both provide coarse (port/queue level) and fine (packet level) granularity that effectively utilize the NIC offload features. As said in the beginning of this chapter, if a software developer does not know the underlying platform that software will run, which is quite in the hardware/software-decoupled design/supplier scenario, some existing approach is to avoid using any NIC offload, which relies on software to do more processing. The other approach is to query the offload capability first and then configure the offload capabilities.

DPDK abstracts the Ethernet device layer, and it is known as `lib/ librte _ ethdev`. This defines the port-/queue-level offload capability in `lib/ librte _ ethdev/librte _ ethdev.h`

```
struct rte_eth_dev_info {
    uint64_t rx_offload_capa;
    /**< All RX offload capabilities including all per-queue
ones */
    uint64_t tx_offload_capa;
    /**< All TX offload capabilities including all per-queue
ones */
    uint64_t rx_queue_offload_capa;
    /**< Device per-queue RX offload capabilities. */
    uint64_t tx_queue_offload_capa;
    /**< Device per-queue TX offload capabilities. */
}
```

When PMD initializes the device, it can announce and report the NIC offload capability. In the example of Intel® 82599 PMD, the code is as follows:

```
drivers/net/ixgbe/ixgbe_ethdev.c

dev_info->rx_queue_offload_capa =
ixgbe_get_rx_queue_offloads(dev);
dev_info->rx_offload_capa =(ixgbe_get_rx_port_offloads(dev)
|dev_info->rx_queue_offload_capa);

dev_info->tx_queue_offload_capa =
ixgbe_get_tx_queue_offloads(dev);
dev_info->tx_offload_capa = ixgbe_get_tx_port_offloads(dev);
```

This leaves the DPDK-based application to configure (enable/disable) the NIC offload feature list. The offload features can be divided into two categories:

- Header parsing: computation and update,
- Segmentation: splitting a large packet into multiple small packets at the TX side

Translating into the computation context, the offload operating is related to the core and memory and cache-related usage. The offload features can save the core-based computation cycles, memory reference latency, memory copy (for protocol insertion) latency, etc.

9.4 HEADER PARSING

9.4.1 VLAN Offload

Most NICs support VLAN off-loading, VLAN is optional, and it is invented to isolate the local networks logically. Since its birth, VLAN has been widely adopted in data centers, enterprises, and broadband access networking. VLAN header is indeed 4 bytes long and often present after the Ethernet header within a packet, and the VLAN header formats are defined below (Figure 9.1):

- VLAN header presence is detected by Ethernet type (TPID) = 0x8100, 2 bytes long
- TCI field is 2 bytes long,
  - 12 bits: VLAN ID (tag), logical network isolation
  - 3 bits: priority, often used for QoS purposes
  - 1 bit: CFI, canonical format indicator (MAC address)

When 4-byte 802.1q tag (VLAN header) is inserted to (or stripped from) an Ethernet packet, the packet integrity is protected by CRC; due to packet content change, the CRC is required to redo; and the result is appended in Ethernet packet trailer. If this VLAN insertion or stripping is done by software, each insertion or stripping results in extra load on the CPU:

- Packet movement; this is translated into the memory access
- CRC-based computation on the whole packet memory

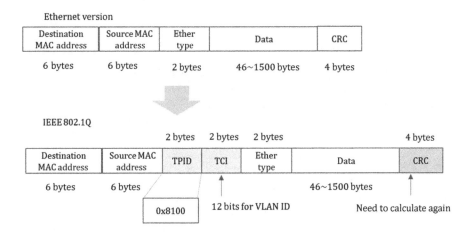

FIGURE 9.1   Ethernet/VLAN packet format.

In the worst-case scenario, if the packet-based memory is not cached yet, the memory access may take hundreds of cycles to get it completed. NIC can do the VLAN off-loading on the NIC, and VLAN insertion and stripping can save the CPU cycles.

It is indeed possible that more than one VLAN header is present in the same packet. Dual VLAN is also known as QinQ VLAN stacking. Some NICs can also off-load the QinQ use case:

1. RX VLAN Filtering

   VLAN filter is configurable on a port level, which receives packets with the same VLAN ID. The packets that do not have the same VLAN ID are not belonging to the same local network, which should be dropped directly. This is one of the typical NIC offload features. It is implemented by using the VLAN filter table on the NIC port. VLAN packets that do not match the filter table will be dropped. There is an example code in `app/test-pmd` on how to set up the VLAN filtering mechanism. The below table provides the command options (Table 9.4).

   The VLAN filtering may cause the packet drop; if this happens, the developer may see no packets are received; and this can be an expected behavior due to VLAN filtering on the port. Such problems should be analyzed by checking the NIC filtering. VLAN filtering at the RX side is just one of the packet filtering mechanisms, and the exact filtering action may vary depending on the actual NIC in use.

2. RX VLAN Stripping

   NIC can remove VLAN header from the received packet. First, the hardware can understand the VLAN header as it parses the packet header. To

TABLE 9.4

VLAN Commands in Test-pmd

Test-pmd Commands	Functions
`vlan set filter (on/off) (port_id)`	Set VLAN filtering on|off on the port. If on, the VLAN packets that failed to match the VLAN filter table will be discarded.
`rx_vlan set (inner|outer) tpid (value) (port_id)`	Set TPID option of VLAN filtering. Support multiple TPIDs.
`rx_vlan add (vlan_id/all) (port_id)`	Add vlan_id to the filter table or all. Multiple VLANs can be added. The maximum number of VLANs supported is subject to the NIC; please refer to the datasheet.
`rx_vlan add (vlan_id) port (port_id) vf (vf_mask)`	Add vlan_id to the filter table on the specified port/VF.
`rx_vlan rm (vlan_id/all) (port_id)`	Remove vlan_id from the filter table on the port, or remove all entries
`rx_vlan rm (vlan_id) port (port_id) vf (vf_mask)`	Remove port/VF and set VLAN filter table

enable this feature, the appropriate bit should be set on the NIC port (or the queue of the NIC port). The NIC hardware will read the configuration register to determine whether VLAN stripping is required on the received packet. DPDK's app/test-pmd shows how to enable and disable the stripping feature on specific port/queue with the commands below:

```
test-pmd> vlan set strip (on|off) (port _ id)
test-pmd> vlan set stripq (on|off) (port _ id, queue _ id)
```

When the stripping feature is enabled, the NIC hardware will strip off a 4-byte VLAN header from the packet. The VLAN header might be useful, so it cannot be discarded. Therefore, the NIC hardware will update two fields in the buffer descriptor:

- VLAN ID (stripped)
- A bit to indicate if strip is completed

The software can read the buffer descriptor to retrieve the stripped VLAN ID. DPDK PMD supports this. If the packet has been stripped of a VLAN tag, DPDK PMD updates the below fields in rte_mbuf for further use:

- The offload flag bit: PKT _ RX _ VLAN _ PKT
- The stripped VLAN: vlan _ tci

```
Struct rte _ mbuf{
uint16 _ t vlan _ tci; /**< VLAN Tag Control Identifier(CPU
order) */
}
```

3. TX VLAN Insertion

Vice versa, it is common to insert a VLAN tag into the packet at the TX port. The VLAN header consists of two parts:

- TPID (tag protocol identifier), which indicates the Ether type of the next header such as IPv4
- TCI (tag control information).

The TPID can be set to a NIC register of the port, and TCI is packet-specific, so the update should be set on a packet basis:

- Write the buffer descriptor rte_mbuf before packet transmission
- Set the offload bit, PKT _ TX _ VLAN _ PKT
- Write the TCI into the vlan _ tci field before the transmission.

```
struct rte _ mbuf{
uint16 _ t vlan _ tci; /**< VLAN Tag Control Identifier(CPU
order) */
}
```

DPDK PMD has the TX function, and SW API will check the buffer descriptor and send command to the NIC, so the NIC can insert the VLAN header before the packet leaves the TX port. Below are the test-pmd commands for VLAN insertion on the TX side. Refer to the code and learn how to take advantage of this feature elsewhere.

```
test-pmd> tx _ vlan set (port _ id) vlan _ id[, vlan _ id _ outer]
```

For instance, to configure all outgoing packets with VLAN ID =5 on port 0, the following command may be used:

```
test-pmd> tx _ vlan set 0 5
```

For instance, to configure all outgoing packets with double VLAN, such as inner VLAN ID = 2 and outer VLAN ID = 3 on port 1, the following command may be used:

```
test-pmd> tx _ vlan set 1 2 3
```

4. Double VLAN

The VLAN technology is very popular and widely used. The VLAN ID is only 12 bits wide so that it can only support only up to 4096 virtual local networks, and this is not enough for large-scale data center network deployment. To address this limitation, double or even multiple VLAN headers (known as VLAN stack mode) are proposed and then adopted, so that Ethernet packets may include multiple VLAN headers before the IPv4/IPv6 header.

This mode (also known as VLAN QinQ) has become common in networking deployment. Many advanced NIC hardware systems are capable of off-loading dual VLAN, i.e., stripping and inserting VLAN for the inner or outer header, and software support is ready in DPDK. Test commands are provided in test-pmd. QinQ has other known names, such as extended VLAN, in some NIC datasheets. The term QinQ is preferred by the DPDK source code. The exact NIC behavior on supporting QinQ may be different, subject to hardware definition.

```
test-pmd> vlan set qinq (on|off) (port _ id)
```

- Take ixgbe PMD (for Intel® 82599 and X550 NIC) as an example: At the packet-receiving side, VLAN filtering and stripping are performed on the inner VLAN tag of double-layer VLAN.
- Take i40e PMD (for Intel® XL710 NIC) as an example: VLAN stripping can remove both the inner and outer VLAN headers. Both VLAN headers can be placed into the packet buffer descriptor in rte _ mbuf.

At the TX side, some NICs may support the inner VLAN insertion only. In this case, the outer VLAN will be owned by software, and the NIC can be under the presence of outer VLAN header, and the inner VLAN insertion can happen in the correct position of the packet header. The NIC behavior can be very specific and different.

9.4.2 IEEE1588 OFFLOAD

IEEE1588 defines the PTP (Precision Time Protocol) for time synchronization. This is a very suitable feature for off-loading. It requires a high standard for real-time performance. The time-stamping operation is expected to have very low latency. This basically rules out software stamping, which is packet modification, and can cause the high latency. Instead, acquiring and writing the time stamp is done quickly in hardware, which can improve the precision. The PTP packet format is defined here (Figure 9.2).

The IEEE1588 PTP basic process can be divided into two phases: initialization and time synchronization (Figure 9.3).

At the initialization phase, each node that is potentially a master node will send a synchronous packet with clock-related information. Meanwhile, each node will receive sync packets sent by other nodes. The nodes will select a master node, based on the Best Master Clock (BMC) algorithm. After selection, the node has its role in the network: either master node or slave node.

After initialization is complete, the time synchronization process will begin. The master and slave nodes have different time synchronization processes.

- The master is required to send sync packets with time stamps in a periodic manner. To increase precision, the time stamp may be added by the NIC hardware as it is closer to the physical layer. The follow-up packet containing the egress time stamp of the sync packet will be sent after the sync packet. Upon receiving a delay request packet, the master node shall record the time stamp and respond with a delayed response packet.
- The slave node needs to record the time stamp of the sync packet sent from each selected master.

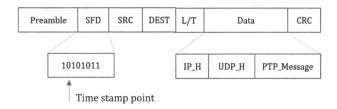

SFD—start of the frame delimiter

SRC—source MAC address

DEST—destination MAC address

L/T—length/type

FIGURE 9.2 IEEE1588 PTP packet format.

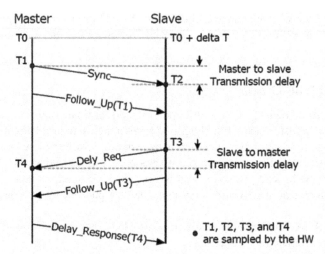

Calculated delta T = [(T2-T1)-(T4-T3)]/2 // assuming symmetric transmission delays
Toffset = - delta T // offset at the slave

FIGURE 9.3   IEEE1588 PTP basic process.

In the above process, the responsibilities are as follows:

- Identifying which packets require time stamps.
- Writing time stamps to those packets which require them. This may be done in both the RX and TX directions.
- Storing the added time stamp in order to inform the software of the time stamp value.

DPDK's app/test-pmd gives an example of how to use IEEE1588, and a forwarding mode for IEEE1588 is supported. IEEE1588 packet transmission and reception are done on a special thread, which uses the NIC off-loading feature. If packets are time-stamped only by software, the time stamp will be regarded as not enough precision because of the software processing. In order to use IEEE1588, the compilation flag needs to be turned on while compiling DPDK:

```
CONFIG_RTE_LIBRTE_IEEE1588=y.
```

Let's take a close look at the following function:

```
int rte_eth_timesync_enable (uint8_t port_id) in DPDK source
code.
```

Using this function, the caller can enable IEEE1588 functionality in both the RX and RX sides. After IEEE1588 is enabled for the specific port, the hardware will be responsible for doing the time stamp on the packets in both directions, and the time stamp can be written into a register and then used by the driver. DPDK provides two functions to read the send and receive time stamps:

```
int rte_eth_timesync_read_rx_timestamp (
uint8_t port_id, struct timespec * timestamp, uint32_t flags)
int rte_eth_timesync_read_tx_timestamp (
uint8_t port_id, struct timespec * timestamp).
```

Although DPDK showcases the basic offload functionality, the example does not include a complete IEEE1588 protocol stack implementation.

9.4.3 CHECKSUM OFFLOAD

Checksum is part of the fault-tolerant design of network protocols. Network transmission is unreliable, and Ethernet, IPv4, UDP, TCP, and SCTP headers all contain their own checksum field (also algorithm) to verify the packet correctness. Checksum calculation is not complicated, but different protocols may have different checksum algorithms, which can be implemented in hardware, as NIC offload feature. Please note that while checksum calculation can be off-loaded to NIC, the NIC still requires the driver software to set up the packet context descriptor in order to use the hardware feature. Checksum field is part of the packet header, and checksum update might be a memory access with high latency.

In the DPDK app/test-pmd application, a checksum forwarding feature is provided for programming reference. Checksum off-loading should be supported in both the sending and receiving directions, but the operations in each direction are different.

In the RX side, NIC will verify the checksum if the port configuration is on, the checksum validation is done in all the incoming packets, the invalid packets can be detected, and the error packets may be dropped and reflected in NIC packet statistics. In DPDK, if error packets are received by PMD, the software driver will update the ol _ flags in rte _ mbuf.

```
PKT_RX_L4_CKSUM_BAD: indicates checksum failed on Layer 4
protocol
PKT_RX_IP_CKSUM_BAD: indicates checksum failed on Layer 3
protocol
```

It is a very simple task for programmers to enable checksum verification on the RX side. It is a different story on the TX side. The NIC needs to be guided on how to do the checksum computation per packet protocol, and the software driver needs to provide the per-packet-based help (to prepare the context descriptor) to assist the NIC offload. The context descriptor should be written into the NIC via PCIe bus, and with that, NIC can perform the checksum computation for the packet to be sent.

The context descriptor should be set up for the TX packet, which has been implemented in the DPDK PMDs. The programmer using DPDK needs to do the following:

- Set up the buffer descriptor such as rte_mbuf
- Prepare for the packet header field

This will guarantee the NIC driver gets enough information to set up the TX context descriptor properly; hence, the NIC can complete the checksum computation and then can update the checksum field on the specified protocol header. The following fields in the rte_mbuf structure are related to checksum off-loading in the sending side:

```
/* fields to support TX offloads */
uint64_t tx_offload; /**< combined for easy fetch */
struct {
        uint64_t l2_len:7; /**< L2 (MAC) Header Length. */
        uint64_t l3_len:9; /**< L3 (IP) Header Length. */
        uint64_t l4_len:8; /**< L4 (TCP/UDP) Header Length. */
        uint64_t tso_segsz:16; /**< TCP TSO segment size */
```

There is no checksum required for the IPv6 packet, and computation is not necessary. For the IPv4 header, the checksum is required, and the NIC can complete the computation and update the field with the following preparations being ready:

```
ipv4_hdr->hdr_checksum = 0; //Clear checksum field of the
header
ol_flags |= PKT_TX_IP_CKSUM; //Set checksum request of IP
layer
```

UDP/TCP protocols have the same checksum algorithm. The following preparations are required to instruct PMD and NIC to work on it:

```
udp_hdr->dgram_cksum = 0; //Clear checksum field of the header
    ol_flags |= PKT_TX_UDP_CKSUM;

//Set checksum request of UDP layer
udp_hdr->dgram_cksum = get_psd_sum(l3_hdr,
info>ethertype,ol_flags);

/* Fill in IP-layer pseudo header computational code. See DPDK
code for its specific implementation. */
```

```
/* SCTP checksum */
sctp_hdr->hdr_checksum = 0;
ol_flags |= PKT_TX_SCTP_CKSUM;
```

In general, checksum computation takes the entire packet content to compute. A packet will have multiple layers of packet header (such as onions), each header has its own checksum field, and all packets are needed to do the checksum in both the RX and TX directions. From the system load level, without NIC offload, the SW-based checksum processing on all packets is considered as a big overhead.

9.4.4 Tunnel Offload

The network overlay technology has emerged in cloud-based networking in the past five years, and VXLAN and NVGRE are the leading examples; these protocols are

FIGURE 9.4   VXLAN packet format.

tunnel-based and add the outer packet headers on the existing TCP/IP-based packet. Network overlay is known as new SDN approach to build the L3 logical networking for large-scale data center. Prior to this, the data center networking is built on L2 (VLAN-based) logical networking.

NIC starts supporting the tunnel-relevant hardware off-loading feature. Different NICs have different features on how to off-load these protocols:

- Some NICs may provide simple checksum offload features only;
- Some NICs may have complex capabilities such as VXLAN and NVGRE packet parsing, which can send traffic to the tenant directly, and also supporting checksum offload for the tunnel protocols—IP and TCP/UDP.
- Some NICs may even support off-loading the addition or stripping of the outer VXLAN and NVGRE header, so that the burden on the CPU can be reduced significantly.

Here is an example of packet header structure of network overlay protocol for VXLAN, and VXLAN adds the outer packet header such as Ethernet header, VLAN header, IP header, and UDP/VXLAN header. The original packet already has its own Ethernet header and packet data, and the original packet is kept intact and encapsulated with the outer header (Figure 9.4).

Now, the new packet (with network overlay feature) has the outer and inner packet headers. The outer header allows the packet going through the data center networking (and even Internet). The inner header is specific to the virtual machines (real workload running as tenants) on the server. The server platform needs to support this overlay protocol, steer the guest traffic to the correct tenant, and add/remove the tunnel headers if required. This is the new norm of networking technology used for the public cloud computing.

Because the NIC features a readiness and development journey, supporting VXLAN and NVGRE tunnel off-loading in DPDK has been an evolving journey. **Test-pmd** has the below support command, and also the code reference on how to enable the offload feature on NIC:

```
test-pmd> port config <port_id> udp_tunnel_port add|rm
vxlan|geneve|vxlan-gpe <udp_port>

test-pmd> rx_vxlan_port add (udp_port) (port_id)
/* Add an UDP port for VXLAN packet filter on a port */

test-pmd> rx_vxlan_port rm (udp_port) (port_id)
```

```
/* Remove an UDP port for VXLAN packet filter on a port */

test-pmd> tunnel_filter add (port_id) (outer_mac) (inner_mac)
(ip_addr)
(inner_vlan) (vxlan|nvgre|ipingre|vxlan-gpe)
(imac-ivlan|imac-ivlan-tenid|
imac-tenid|imac|omac-imac-tenid|oip|iip) (tenant_id)
(queue_id)
/* add a tunnel filter of a port */

test-pmd> tunnel_filter rm (port_id) (outer_mac) (inner_mac)
(ip_addr)
(inner_vlan) (vxlan|nvgre|ipingre|vxlan-gpe)
(imac-ivlan|imac-ivlan-tenid|
imac-tenid|imac|omac-imac-tenid|oip|iip) (tenant_id)
(queue_id)
/* remove a tunnel filter of a port */
```

9.4.5   SEGMENTATION OFFLOAD (TSO)

TCP supports the packet segmentation, which is designed to send the large application data through Internet. Some Internet middle network box may only support the limited size of packet routing, so the segmentation must be done at some scenario. If the NIC can off-load this segmentation at the TX side, it is known as TSO. The NIC can significantly reduce the CPU burden using TSO, which moves the packet from one memory to multiple memories, illustrated in the figure below. In addition to the memory copy, all new packet headers need to be constructed on the original packet header, update the length-related information in the header, and recalculate the checksum (Figure 9.5).

DPDK's app/test-pmd supports how to enable TSO on a port:

```
test-pmd> tso set (segsize) (portid) /* config seg size on the
port */
test-pmd> tso show (portid) /* show tso status on the port */
```

After turning on the port configuration, the software needs to set up the context descriptor so that each packet memory buffer (rte_mbuf) has the instruction to PMD and NIC to enable TSO accordingly.

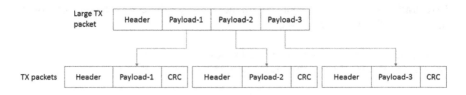

FIGURE 9.5   TCP packet segmentation.

```
/* fields to support TX offloads */
uint64_t tx_offload; /**< combined for easy fetch */
struct {
      uint64_t l2_len:7; /**< L2 (MAC) Header Length. */
      uint64_t l3_len:9; /**< L3 (IP) Header Length. */
      uint64_t l4_len:8; /**< L4 (TCP/UDP) Header Length. */
      uint64_t tso_segsz:16; /**< TCP TSO segment size */
}
```

A bit set is required on ol _ flags: PKT _ TX _ TCP _ SEG, which will guide NIC PMD (which guides the NIC) to enable TSO on this packet. It is worth mentioning that the TSO can become complicated together with the use of network overlay protocol (VXLAN), and more description and the code example are given in test-pmd/csumonly.c.

9.5 CONCLUSION

Hardware acceleration is an interesting topic in the platform context, and the platform usually includes the general-purpose CPU, chipset, or a variety of interface cards. NIC is the ingress hardware for packet reception, and it is the egress hardware for packet transmission. NIC offload can reduce the processing load of the CPU. It is a hardware benefit. But to be clear, NICs can be very different in terms of supported offload feature, and some NICs may support VXLAN off-loading while other hardware may not. Different vendors may implement different VXLAN offloads, and the exact feature can be different.

The software developers are required to have a thorough understanding before using any NIC offload features and learn the details on what are available and then how to use them. DPDK has supported lots of NIC offload in software abstraction, and it is important to look into them under a specific NIC context. There is a common case that software designer cannot predict the underlying platform details, and DPDK provides APIs to query the common set of NIC offload features. When we use the offload feature, it will definitely bring performance benefits.

10 Packet Security

Fan Zhang (Roy) and Heqing Zhu
Intel®

CONTENTS

Security processing such as data encryption, integrity check, and data compression is a foundational measure for Internet network security (in transit) and data storage (at rest). Encryption everywhere is the new recommended security practice for securing data in transit and at rest. Packet security is often implemented by protocols such as IPsec, SSL/TLS, SRTP, QUIC, and WireGuard. Regardless of a specific protocol, security processing has the following key aspects:

- Data confidentiality: The data is encrypted in the source node and decrypted at the destination node, so that the data is secured for Internet transit. This encryption is typically achieved by symmetric crypto-algorithms such as AES and ChaChaPoly.
- Data integrity: The packet should not be modified during its transmission from the source to the destination. The integrity check uses the typical algorithms such as SHA. Encryption was expensive computation to use

CPU cycles, integrity check can happen earlier, and if the packet integrity is corrupted, it should be dropped without using the decryption.
- Secret key exchange: The network source and destination nodes use public/private key pair to derive the secret keys for data encryption/decryption. This is known as asymmetric crypto, and the known algorithms are RSA, Elliptic Curve, etc.

10.1 EARLY SOFTWARE-BASED CRYPTOGRAPHY

As said above, the crypto-computing does not come cheap. Take pure software implementation of computing ciphertext using the AES-128 algorithm (Daemen, Borg, & Rijmen, 2002) as an example: The 128-bit plaintext is sliced into a 4×4 block as input. To encrypt the block, a definite ten rounds of operations are executed. In each round, the block needs to be XORed with a block of pre-computed round key, followed by a nonlinear lookup-and-replace operation and several row shifting operations. In the end, the block's columns are mixed as the output of next round input In (More & Bansode, 2015). Such a complex procedure was a big overhead to the CPU. Schneier, Kelsey, & Doug (1999) stated that for general-purpose CPUs such as Intel® Pentium Pro processor, the encryption of AES takes 18 cycles per byte, equivalent to 0.144 ms per block, or 11 MB/s for a 200-MHz processor. Pure software and sequential implementation of cryptographic operation is definitely not what traditional general-purpose CPUs are good at.

10.2 INTEL® AES-NI AND MULTI-BUFFER

In 2010, Intel® released AES New Instructions (AES-NI) in the "Westmere" processor. Since then, AES-NI has been available on the x86 processor. There are six instructions that offer full hardware support for AES. Four instructions support the AES encryption and decryption, and two instructions support the AES key expansion. AES-like cryptographic algorithms use the repeated operations for multiple data blocks. This suits well to single instruction, multiple data (SIMD) to process multiple data blocks in one operation. In today's mainstream Xeon server, AES-GCM is able to achieve 0.64 cycles per byte. For the next-generation server (Ice Lake), it introduced the additional vector AES instructions, and it is possible to push the performance limit to 0.16 cycles per byte, https://eprint.iacr.org/2018/392.pdf.

For the algorithms that cannot be implemented in SIMD, such as counter mode (CTR) or CBC where the next block computation depends on the output of current block computation output, the software optimization idea is to introduce the *multi-buffer* approach to reduce this data dependency to achieve ideal performance. The idea of multi-buffer technology is as follows: Since every round in the AES implementation is the same (except for one extra XOR in the first round and no column mix in the last round), multiple independent buffers can be encrypted in parallel, so that one *AESENC* instruction can be applied to these buffers simultaneously.

Use IPsec as an example. *Intel® published the open-source multi-buffer IPsec library.* It supports multiple symmetric algorithms implemented by Intel® SIMD intrinsics and multi-buffer technology, and this library has been integrated with the

DPDK Cryptodev library. This software is CPU-based implementation, and it can be used without DPDK.

10.3 INTEL® QUICKASSIST TECHNOLOGY (QAT)

QAT is the Intel® accelerator for crypto-processing and compression processing. It aims to free up the computation burden from the CPU and is shown as a PCIe device in the server system. QAT silicon is available on a PCIe add-in card, or a specific Intel® server chipset, or Intel® SoC. The software driver will run at the CPU, and it will prepare and send the crypto-requests to the QAT device. QAT delivers the encrypted packet back via DMA and response message. In the meantime, CPU can work on other tasks before the data encryption is finished, so this is an asynchronous model. QAT driver support has been integrated into open-source projects such as Linux kernel, OpenSSL, and FD.io/VPP. DPDK Cryptodev also integrates the QAT support.

10.4 DPDK CRYPTODEV

DPDK Cryptodev is a set of APIs as the generic interface, it gains the support from other silicon vendors, and it has gained the broad industry support with multiple-vendor driver integration. The Cryptodev goal is to design the software abstraction layer, and it works with software (CPU)- or hardware-accelerated crypto-devices. The Cryptodev can hide the underlying silicon differences to "DPDK user space application". The Cryptodev concept is illustrated below (Figure 10.1).

The Cryptodev API has the following features:

- Device management API is used to configure the number of queue pairs and NUMA awareness. Each queue pair, as the name indicates, refers to two queues one for the application to enqueue the crypto-operations to be processed, and the other for the application to dequeue from the processed crypto-operations. Queue pairs are accelerator-specific. For the software crypto such as AESNI-Multi-Buffer, queue pairs are realized by system memory.
- Queue pair management API is responsible for memory allocation and initialization used by the user, such as `rte _ cryptodev _ queue _ pair _ setup`.
- Device capabilities API is designed to report the crypto-capability, such as the supported crypto-algorithms and its details such as digest lengths. A data structure `rte _ cryptodev _ capabilities` is defined so that the device can declare all the supported capabilities. The user application can query `rte _ cryptodev _ sym _ capability _ get` or `rte _ cryptodev _ asym _ capability _ get` to obtain the support symmetric or asymmetric algorithms, key lengths, digest lengths, etc.
- Session management API is the crypto-session relevant abstraction layer, which supports the common and device-specific data area for temporary keys, initial vector and AAD lengths, etc.
- Enqueue and dequeue APIs are burst-based: `rte _ cryptodev _ enqueue _ burst` / `rte _ cryptodev _ dequeue _ burst`. In

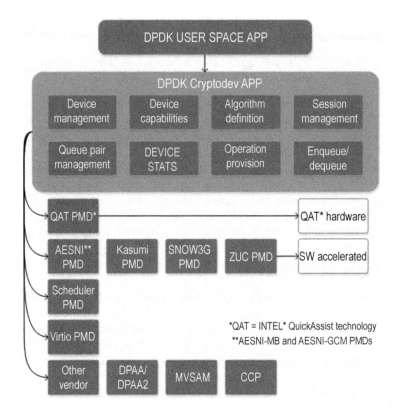

FIGURE 10.1 DPDK Cryptodev framework structure.

addition to the rte_mbuf, rte _ crypto _ op is introduced to support the security session, the data offset to start encryption/authentication, and the virtual and physical addresses of the digest and AAD.

- Synchrnous CPU crypto processing API: rte _ cryptodev _ sym _ cpu _ crypto _ process. The API is designed for software crypto drivers such as Intel® AES-NI PMD to process crypto workload synchronously (instantly return the processed workloads). The API directly uses data pointers to pass source and destination buffers, digests, and even scatter-gather list buffer arrays, and is independent to rte _ crypto _ op and rte _ mbuf.

10.4.1 Cryptodev PMD

Multiple silicon vendors released the Cryptodev PMDs in DPDK.

- Intel® released the QAT PMD.
- Intel® released a list of SW crypto-PMDs such as AES-NI, Kasumi, ZUC, and SNOW3G.
- Other silicon vendors such as Marvell released their own PMDs.

In order to use Cryptodev, a few things need to be checked:

- Device capabilities: Different PMDs may support different cryptographic algorithms and key sizes. It is important to query the device capabilities before using it, for example, whether a specific cryptographic method is supported.
- `rte _ cryptodev _ ops`: A `rte _ cryptodev _ ops` data structure contains a set of callback function pointers, and it can help the user application to configure the device, queue pairs, and sessions and to get and clear the statistics.
- Enqueue and dequeue callback functions: These two functions are the main datapath operations of Cryptodev PMD. As we mentioned earlier, DPDK takes packets in bursts. The burst mode allows submitting more jobs into the cryptographic hardware in one request; typically, this operation is asynchronous; and CPU and crypto-accelerator are able to work in parallel.

To create a device, the PMD should use the `RTE _ PMD _ REGISTER _ CRYPTO _ DRIVER` to register the driver such as the following:

```
RTE_PMD_REGISTER_CRYPTO_DRIVER(crypto_drv, drv, driver_id)
crypto_drv: is the created driver structure;
drv: a rte_driver structure for hardware PMD, or a rte_vdev_
driver virtual driver for software PMD.
```

An important function set is the probe and remove callback functions to be invoked at the phase of device creation and deletion stages. Cryptodev PMD shall implement both functions. Regarding the probe function, a number of device initializations should be implemented:

- Providing `rte _ cryptodev _ pmd _ init _ params` structure data with the name of the driver, the socket ID number, and the driver's private data size.
- Passing the `init _ params` data pointer to `rte _ cryptodev _ pmd _ create` function to create a PMD. The successful execution of the function will:
  - Create `rte _ cryptodev` structure data that contains a `dev _ priv` pointer for the driver to store its private data, with the size specified in the *init* parameter.
  - Insert the `rte _ cryptodev` structure data into `rte _ cryptodev _ globals` array which contains all existing device data.
  - Return the pointer of the `rte _ cryptodev` structure data to the driver.
- The driver should now initialize `dev _ priv` content, plus the `rte _ cryptodev` structure with all necessary data including driver ID, enqueue and dequeue function pointers, and the `dev _ ops` structure pointer as we mentioned above.

- The driver should also provide a 64-bit unsigned integer with supported feature flag set. All possible features are defined in `rte_cryptodev.h` and there are 24 feature flags available,, each flag indicates a specific driver capability such as `RTE_CRYPTODEV_FF_SYMMETRIC_CRYPTO`, indicating the device supporting the symmetric cryptographic algorithms and `RTE_CRYPTODEV_FF_SYM_OPERATION_CHAINING`, inferring operation chaining such as cipher then hash, or not, by the driver.

10.4.2 How to Use

This section will introduce how to write a simple cryptographic application using the symmetric AES-CBC cipher algorithm and SHA1 hash algorithm to encrypt and authenticate the packets step-by-step. Following the introduction, some tricky points will be discussed. Also please bear in mind that the APIs used are based on DPDK Release 18.08. There might be some code differences in the later DPDK release. The main ideas shall be similar.

10.4.2.1 Cryptodev Assignment

Similiar to other physical or virtual devices, Cryptodev shall be presented to the application, and it can be done with the DPDK EAL command option. The maximum number of VFs can be different to the specific crypto device; for example, DH895XCC QAT supports up to 16 VFs.

QAT is used as an example here, it needs to be assigned to DPDK, so that the Linux kernel driver will release the control.

```
>>./dpdk/usertools/dpdk-devbind.py -s
0000:88:00.0 'DH895XCC Series QAT 0435' if= drv=dh895xcc
unused=qat_dh895xcc
```

where the PCI address `0000:88:00.0` is referring to the QAT device DH895XCC.

DPDK will take QAT VF as the crypto-device, and the next step can configure the number of VFs on a given QAT device.

The below command assigns two VFs to the QAT device DH895XCC:

```
>> echo 2 > /sys/bus/pci/drivers/dh895xcc/0000\:88\:00.0/
sriov_numvfs
```

Expose the Cryptodev to the target application, and use `l2fwd-crypto` as an example.

```
>>./l2fwd-crypto -l 11-n 4--socket-mem 0,2048-w 88:01.0-w
88:01.1-w 84:00.0-- …
```

The PCI device 84:00.0 is the PCI address of QAT VF.

To use software Cryptodev, such as Intel® AESNI-MB PMD, the configuration is different, and as AESNI-PMD is a virtual device, it can start with the "`--vdev`" option.

If you want to use Intel® AESNI-MB PMD, the EAL command option is given below:

```
>> ./l2fwd-crypto -l 11-n 4--socket-mem 0,2048-w 88:01.0-w
88:01.1--vdev "crypto_aesni_mb0" -- …
```

The Cryptodev library uses a global array to store all detected crypto-devices. Each device is assigned with an ID. It starts from 0 with the maximum number of devices defined in `RTE _ CRYPTO _ MAX _ DEVS`. The ID is used for the application to access the device. The DPDK Cryptodev framework takes the hardware crypto-accelerator with higher priority; thus, all software Cryptodev PMDs, if presented along with the hardware ones, will be listed after them. The following API can retrieve the device ID:

```
extern int rte_cryptodev_get_dev_id(const char *name);
```

10.4.2.2   Cryptodev Configuration

The device configuration is simple. What needs to be configured are two parameters, one is the `socket _ id`, which is related to the memory lies upon in a dual-socket system, and the second is the maximum number of queue pairs of the device. The setup of queue pairs is more complex. First, each queue pair should be configured individually. So far, there are two parameters that should be passed into the API `rte _ cryptodev _ queue _ pair _ setup()`. They are the queue size and the session mempool pointer for session-less operation. The size of the queue pair decides how many jobs can be enqueued at the same time. As we mentioned earlier, the operation in DPDK Cryptodev is asynchronous, which means the operations submitted to the Cryptodev are not processed immediately; instead, they are stored in the ingress queue of the queue pair. Also, the Cryptodev-processed job is stored in the egress queue of the queue pair, too. Another important factor is that, unless the latency really matters, the frequency of enqueue operation to the hardware Cryptodev should be minimized to increase the overall throughput performance. We will talk about the reason of it later. The bigger the queue pair size is, the more operations can be queued and processed in a batch, and the bigger the memory footprint of the application. However, using less number of queue pairs reduces the application's memory usage, but the ingress or egress queue will get full soon, which will reject the upcoming operations to be enqueued. For this reason, the queue pair size is an important balance point.

10.4.2.3   Configuring the Session

As we mentioned earlier, the session is used to store expanded key or other intermediate data that can be reused by multiple Cryptodev operations. In DPDK Cryptodev design, the session shall be maintained by the user application. Different Cryptodev PMDs have different format of session data with various sizes. DPDK Cryptodev requires the user to provide one or more mempool pointers as a parameter to create and initialize the session. Since different Cryptodev PMDs have different session data size, your mempool must have the element size equal to or bigger than the PMD's session data size you are using. If you plan to make the same session mempool shared with more than one PMD in your system, the mempool's element size should be equal to or bigger than the biggest session data size of all PMDs. The size of the PMD can be known by `rte _ cryptodev _ sym _ get _ private _ session _ size()` or `rte _ cryptodev _ asym _ get _ private _ session _ size()`.

To make the same session data to be processed by different Cryptodev PMDs, the DPDK Cryptodev framework uses an array of pointers to store multiple PMD-specific session data. The DPDK Cryptodev session structure is shown as follows:

```
struct rte_cryptodev_sym_session {
        __extension__ void *sess_private_data[0];
        /**< Private symmetric session material */
};
```

The above session structure is designed for symmetric cryptographic algorithms, but the asymmetric algorithm has the structure with the same format. Upon application initialization, DPDK Cryptodev will scan all Cryptodev PMDs that exist, and then assign a driver ID for each PMD type. To initialize a session for the operation, the user shall call `rte _ cryptodev _ sym/ _ asym _ session _ create()`. This API requires only one parameter, the mempool pointer to fetch one `rte _ cryptodev _ sym/ _ asym _ session` type object. Depending on the number of PMD types in the application, the function will extend sess_private_data array to match this number, so each PMD has its individual session private data pointer field. However, the session is not ready yet. The user has to call the `rte _ cryptodev _ sym _ / _ asym _ session _ init()` function to initialize the session. Let's take a look at the API to initialize a symmetric algorithm session.

```
int rte_cryptodev_sym_session_init(uint8_t dev_id,
                    struct rte_cryptodev_sym_session *sess,
                    struct rte_crypto_sym_xform *xforms,
                    struct rte_mempool *mempool);
```

Let's assume the dev_id is the Cryptodev PMD's ID, which is the Nth type (which starts from 0) device in the application. Along with dev_id, the user also needs to pass the session created earlier and the symmetric algorithm transform operations into the function. More importantly, the user has to pass a mempool object a second time. This is because the function will allocate another object from the mempool and initialize this data as to the PMD's favorite session data format. In the end, the function lets the sess_private_data [N] in the session points to this object. To make the session available to a second Cryptodev PMD, the user can call the init function again, with the same sess parameter but different dev_id. To free a session back to the mempool, the user needs to call the clear function to free all the elements in the sess_private_data array before calling the free function.

10.4.2.4   Let's Assemble Your Cryptographic Operations

Once the Cryptodev is configured and the queues are set up, and the sessions are created for the specific algorithm with secret keys, it is ready to secure the packets.

In DPDK, the packet data is stored in the `rte _ mbuf`, so as the DPDK Cryptodev will deal with this data structure. In fact, Cryptodev has a bit extra: The mbufs are wrapped with an mbuf, an extra layer is introduced: the cryptographic operation structure `rte _ crypto _ op`, and `rte _ crypto _ op` structure is the unit recognizable by the DPDK Cryptodev.

To make the rte_crypto_op structure compatible with both symmetric and asymmetric Cryptodev PMDs, it contains a union element with both symmetric cryptographic operation (rte_crypto_sym_op) and asymmetric cryptographic operation (rte_crypto_asym_op) arrays, both with zero length elements. It is expected to provide a memory-contiguous buffer, both virtually and physically, to store one rte_crypto_op structure and one symmetric or asymmetric operation structure next to it. From this point, we will focus on symmetric cryptographic operation. A cryptographic operation with symmetric cryptographic operation is shown below (Figure 10.2).

Currently, rte_crypto_op takes one cacheline and rte_crypto_sym_op takes another cacheline. However, we need some extra space to store the initial vector (IV). Upon creating a symmetric cryptographic session, the user needs to set the offset value from the start of rte_crypto_op structure to the IV data, so the Cryptodev PMD can retrieve the IV from the received cryptographic operation. To minimize the memory footprint, the IV should be right next to the rte_crypto_sym_op, so the IV offset should equal to the size of rte_crypto_op and rte_crypto_sym_op combined.

In addition, all fields in rte_crypto_op and rte_crypto_sym_op should be set properly before enqueuing to the DPDK Cryptodev PMD. The fields in the rte_crypto_op structure are simpler, and the fields to be specified include the operation type (symmetric or asymmetric) and session type (with-session or session-less). If the structure is allocated by the rte_crypto_op_alloc() or rte_crypto_op_alloc_bulk() function call, some fields including the mempool pointer or the operation's physical address are set already. Some applications may use the mbuf headroom to store the cryptographic operation, as it helps reduce the overhead of allocating and freeing the cryptographic operation cost. However, since symmetric cryptographic operations

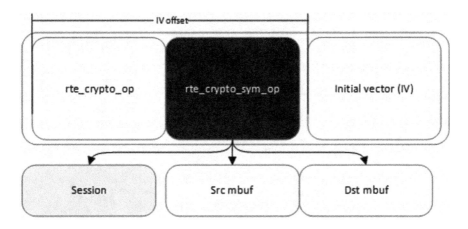

FIGURE 10.2   DPDK crypto-operation structure.

already take 128 bytes of space, the mbuf headroom has to be expanded with extra space to store the IV. Moreover, some fields inside the rte_crypto_op structure, such as the physical address of the operation, need to be carefully calculated and assigned based on the physical address of the mbuf dataroom.

No matter the cryptographic operation is allocated or taken from the mbuf headroom, it is important to know all physical address fields inside the operation structure is for hardware crypto device only. Hardware crypto device (like Intel® QAT) has the corresponding Cryptodev PMDs , the hardware device access the data in memory using DMA approach, and data is being processed further by the hardware device. For software based Crypto PMDs, it is okay to not care about it, it does not see any physical address. If the hardware device-based PMD is used, the PMD has taken care of the physical addresses into consideration. But there is one exception, the authentication and AEAD algorithms' digest data and AEAD's additional authentication data (AAD). If these data are in the packets, one useful suggestion is to compute them by using the APIs provided by DPDK such as rte_pktmbuf_mtod_iova_offset() to obtain the virtual or physical addresses of certain offset in the mbuf.

10.4.2.5   Finally, Enqueue and Dequeue Operations

The same as other DPDK applications, enqueue and dequeue operations are the main datapath APIs in the DPDK Cryptodev framework. A simplified DPDK Cryptodev workflow to secure the network is shown in Figure 10.3. The user needs to implement

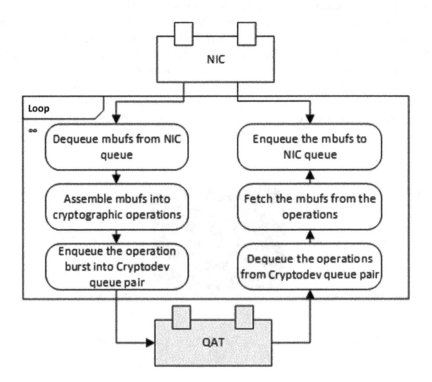

FIGURE 10.3   Datapath workflow with DPDK Cryptodev.

the parts including assembling the mbufs into cryptographic operations, enqueuing to the target queue of the DPDK Cryptodev PMD such as QAT, and then dequeuing from the queue and send the mbufs in the dequeued operations to NICs. DPDK Cryptodev PMD will update the status field in the cryptographic operation. The status with the RTE_CRYPTO_OP_STATUS_SUCCESS flag means the processing is finished successfully. The failed status flags also can be a good reference for the user to decide how to deal with the packets. For example, the flag RTE_CRYPTO_OP_STATUS_AUTH_FAILED may mean the packet comes from an unauthorized party, the packets should then be dropped, and blacklisting the traffic from the packet's IP in the firewall can be the next step.

There is not much left to mention here, but two very important ones, all relating to QAT PMD:

- Bear in mind that "kicking" the QAT to start processing the operations—in other words, an enqueue burst call to the QAT PMD—costs several hundred cycles. As we mentioned above, this call should be minimized as long as the queue pair size is big enough and the latency requirement is not critical for the application.
- Intel® QAT PMD does not have infinite processing capabilities. Each model of QAT hardware has its maximum throughput and varies depending on the packet sizes. When you are tuning your system, you must know these limitations to determine if the QAT capacity is the bottleneck of your system.

10.5   SECURITY LIBRARY

Cryptodev has been recognized as an efficient crypto-workload acceleration engine framework and has been widely used in many security applications. However, the Cryptodev framework is too generic, and it aims to support all popular block and stream cipher, authentication, and AEAD algorithms supported by different SW and HW lookaside crypto-accelerators from different vendors. This introduces some unavoidable overhead to some PMDs to maintain the same behavior as the others. In the meantime, the asynchronous enqueue/dequeue working mode of Cryptodev benefits the most to the lookaside crypto-HW that works in parallel with CPU, which does not fit the synchronous in-line acceleration method.

To improve the situation, the DPDK Security library was born. The DPDK Security library aims to accelerate security protocol processing that works in in-line or synchronous mode. The library defines the generic APIs to create and destroy the security sessions, it supports the datapath packets is fully or partially offloaded to in-line crypto- to hardware-based devices. To date, a few security protocols are supported by security library: packet data convergence protocol (PDCP), IEEE MAC Security standard (MACsec), and IPsec. Moreover, the in-line and synchronous crypto-processing are supported, too.

Figure 10.4 outlines the basic concept of DPDK security library. If the NIC or the crypto-device (or other devices in the future) supports accelerating certain workload such as crypto-processing or protocol processing, the security library provides the user with the APIs to create the security session(s) containing necessary information

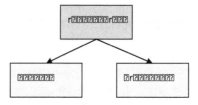

FIGURE 10.4   Security layers and framework.

for the PMD to recognize the workload and process accordingly. If the target device is a NIC, the cooperation of DPDK security library and flow library may be necessary for the NIC PMD to understand the security session being applied to specific flow(s) going through the specific port(s). A specific flow action type, "rte_flow_action_security", is used to pass the security session into the NIC PMD.

A new type of rte_security_action type, "cpu_crypto", is introduced. Different than the NIC-based security action, this action type uses the CPU to process the crypto-workload synchronously. Although already supported in Cryptodev, CPU instruction-accelerated crypto-processing such as Intel® AES-NI instructions does not require the asynchronous enqueue/dequeue working mode and can be processed in-line with the rest of packet processing. To maintain the same behavior as the lookaside hardware such as Intel® QAT, it uses the software queues to simulate the hardware queue structure and enqueue and dequeue operations. Moreover, it has to fill a complex DPDK crypto-operation structure to describe the workload. The cpu_crypto security action support in DPDK AESNI-MB and AESNI-GCM PMDs removes these bottlenecks to improve the crypto-workload processing performance.

10.6   IPSEC LIBRARY

With the Cryptodev and security libraries, we now have two distinctive ways to process crypto-workload, each with different supported crypto-devices, software, lookaside hardware, and in-line hardware. In terms of IPsec datapath workload, crypto is only part of the job. The IPsec gateway application still needs to identify the flows and lookup the associated SA (Security Association), encap/decap tunnel headers, and route the packets to the specific ports. Is there an easier way to put all these tasks together?

The DPDK IPsec library is proposed to include all dots together. It is a modular library built around a core functionality of datapath processing and SA management. Moreover, it supports all different software/hardware/in-line crypto-devices' acceleration into the same API. The library automatically handles HW accelerator allocation and resource usage. In the future, the optional modules including SAD, SPD, crypto-load balancing, and IKE shim layer are to be added to the library (Figure 10.5).

The IPsec library also adopts the session idea similar to the DPDK Cryptodev and security libraries. The below figure shows the process of creating an IPsec session. The same DPDK Cryptodev xform structure is used by both Cryptodev and security devices to pass crypto-parameters. Since it supports both libraries, either of their sessions can be passed into the IPsec library to create an IPsec session (Figure 10.6).

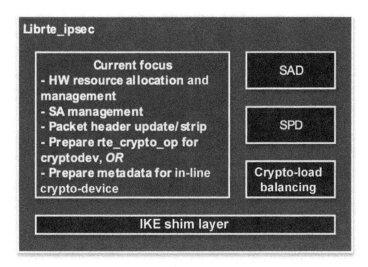

FIGURE 10.5   IPsec library.

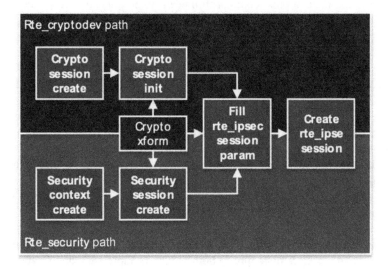

FIGURE 10.6   IPsec session creation.

The below figure shows the DPDK IPsec library datapath workflow. Thanks to the IPsec session recorded information, different crypto-processing patch will be chosen automatically based on the IPsec session action type (Figure 10.7).

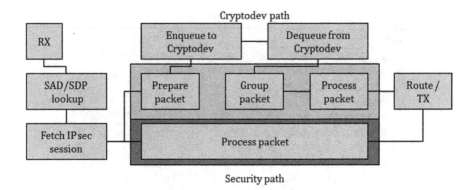

FIGURE 10.7   IPsec datapath.

For those who want to know the latest performane, Intel published the crypto and IPsec performance metrics here, it is updated in a quarterly basis. https://fast.dpdk.org/doc/perf/DPDK_20_05_Intel_crypto_performance_report.pdf

10.7   NETWORK COMPRESSION

Data compression is a method to reduce the data size, but compression does not come free, and the compression algorithm needs to be implemented by the CPU or accelerator.

10.7.1   COMPRESSDEV FRAMEWORK

DPDK Compressdev was a new feature in early 2018. DPDK Compressdev is a collaboration effort contributed by Intel®, Cavium™, and NXP Semiconductors. Followed by the design ideas similar to DPDK Cryptodev, DPDK Compressdev also provides a DPDK Compress Device Library and a number of compression PMDs for software/hardware compression accelerations. The Compressdev focuses on accelerating the network and storage compression. DPDK Compressdev is designed to use the unified API for both use cases (Figure 10.8).

The structure of DPDK Compressdev is illustrated in the below figure, and the similarities between DPDK Compressdev and Cryptodev are as follows:

- Asynchronous burst API. All performance benefits and impacts of accessing hardware accelerators apply to both DPDK Compressdev and Cryptodev. The asynchronous operation helps the accelerator to compress the data in parallel, and the burst operation helps averaging the cost of "kicking" the hardware to all packets.
- Use mbufs as data carrier. Mbuf currently has the maximum data size of 64K bytes minus one. To process the data bigger than this size, the user can use chained mbufs.
- Multiple queues for CPU threads sharing the same Compressdev resource.
- Software and hardware PMDs sharing the same API for effortless migration.

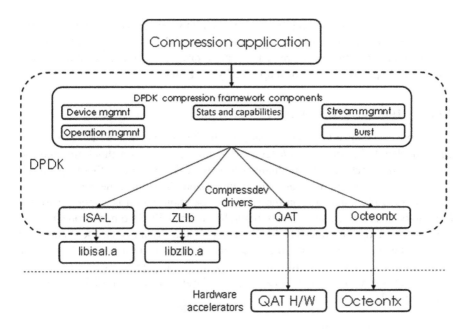

FIGURE 10.8 DPDK Compressdev components.

So far, DPDK Compressdev contains four compression PMDs: two in software—Intel®
ISA-L PMD that employs the compression engine from Intel® ISA-L library that is
optimized for Intel® Architecture, and Zlib PMD that enwraps the Zlib compression
library; and two in hardware—Intel® QAT PMD that utilizes Intel® QuickAssist
family of hardware accelerators, and Cavium Octeontx PMD that uses HW offload
device found in Cavium's Octeontx SoC family.

10.7.2 The private _ xform for Stateless Operation

The `private _ xform` data structure is used to store the Compressdev PMD's
private data for processing the operations' stateless compression. The same as the
DPDK Cryptodev cryptographic session, the private_xform instance is designed to
be maintained by the user application. To create private_xform data, the user has
to fill a `rte _ comp _ xform` structure with the `op _ type` defined as `RTE _
COMP _ OP _ STATELESS`. The structure contains the information needed by the
PMD to initialize the opaque data, such as operation of compression/decompression,
compression algorithm (so far, only the DEFLATE algorithm is supported), algorithm parameters, window size, compression level, and hash and CRC generation/
verification enable/disable. Once the structure is filled, it can be passed to the API as
shown below to create the private_xform:

```
int __rte_experimental
rte_compressdev_private_xform_create(uint8_t dev_id,
          const struct rte_comp_xform *xform,
          void **private_xform);
```

The created private_xform can then be attached into compression operations before enqueuing to the PMD. Different than the Cryptodev's session shareable between cryptographic operations, the created private_xform may not be shareable between compression operations if is not supported by the PMD.

10.7.3 The Stream for Stateful Operation

One compression operation has the data size limitation, which is 64 KB minus one. Once the data to be compressed/decompressed has the size over this limit, multiple compression operations can be chained to extend the range and these operations are treated as a data stream. For many compression algorithms, the compression of one operation often depends on the previous operation, and it is important for the PMD to be aware of the data stream beforehand. The DPDK compression stream is used to do exactly the same. The idea of compression stream is used to define a set of operations in a data stream. To create a data stream, the user has to fill the same rte _ comp _ xform structure with the op _ type defined as rte _ comp _ op _ type. Once the xform is filled properly, the stream can be initialized with the following API:

```
int __rte_experimental
rte_compressdev_stream_create(uint8_t dev_id,
          const struct rte_comp_xform *xform,
          void **stream);
```

10.7.4 Enqueue and Dequeue Operations

The APIs are used to submit and retrieve the compression operations. DPDK Compressdev supports compression/CRC/hash in single operation, and the status bit in the compression operations will not be set as RTE _ COMP _ OP _ STATUS _ SUCCESS unless all operations are completed successfully.

```
uint16_t __rte_experimental
rte_compressdev_enqueue_burst(uint8_t dev_id, uint16_t qp_id,
          struct rte_comp_op **ops, uint16_t nb_ops);

uint16_t __rte_experimental
rte_compressdev_dequeue_burst(uint8_t dev_id, uint16_t qp_id,
          struct rte_comp_op **ops, uint16_t nb_ops);
```

REFERENCES

Beckers, K. (2015). *In Pattern and Security Requirements: Engineering-Based Establishment of Security Standards* (p. 100). Springer, Cham.
Cisco. (2017, June 06). *Cisco Visual Networking Index: Forecast and Methodology, 2016–2021.* Retrieved from https://www.cisco.com/c/en/us/solutions/collateral/service-provider/visual-networking-index-vni/complete-white-paper-c11-481360.pdf.
Daemen, J., Borg, S., & Rijmen, V. (2002). *The Design of Rijndael: AES - The Advanced Encryption Standard.* Springer, Berlin.

Duman, J. (2017). *FD.io VPP - whitepaper*. Cisco.

Frankel, S., Glenn, R., & Kelly, S. (2003). *The AES-CBC Cipher Algorithm and Its Use with IPsec*. Retrieved from RFC: http://www.rfc-editor.org/info/rfc3602.

Gopal, V. (2010). Processing Multiple Buffers in Parallel to Increase Performance on Intel® Architecture Processors. Intel®.

Guilford, J., Gulley, S., & Erdinc, O. (2012). Fast Multi-buffer IPsec Implementations on Intel® Architecture Processors. Intel®.

Intel® DPDK Validation Team. (2017). *DPDK Intel® Cryptodev Performance Report*. Intel®, Shanghai. Retrieved from https://fast.dpdk.org/doc/perf/DPDK_17_11_Intel_crypto_performance_report.pdf.

More, S., & Bansode, R. (2015). Implementation of AES with time complexity measurement for various input. *Global Journal of Computer Science and Technology: ENetwork, Web & Security*, 15(4), 10–20.

Patterson, D., & Hennessy, J. (2013). *Computer Organization and Design MIPS Edition: The Hardware/Software Interface*. Newnes, Oxford.

Schneier, B., Kelsey, J., & Doug, W. (1999). Performance comparisons of the AES submissions. *Second AES Candidate Conference* (pp. 15–34). NIST.

Tsirkin, M.S. (2016, March 03). *Virtual I/O Device (VIRTIO) Version 1.0*. Retrieved from Virtio: http://docs.oasis-open.org/virtio/virtio/v1.0/virtio-v1.0.pdf.

Viega, J., Messier, M., & Pravir, C. (2002). *Network Security with openSSL: Cryptography for Secure Communications*. O'Reilly, Sebastopol, CA.

https://www.intel.com/content/dam/doc/white-paper/advanced-encryption-standard-new-instructions-set-paper.pdf

Section 2

I/O Virtualization

With the rise of virtualization technology, computing workload can run in the virtual machine (VM) with its own OS; this allows multiple workloads (even from different organizations) to be consolidated in the same physical machine; this model transforms into the cloud computing; it meets the law of economics. Super cloud service providers (AWS – Amazon Web Service, Azure Cloud, etc.) emerged as the leading technology force to support today's IT infrastructure and service (PaaS, IaaS, and SaaS).

Consolidating the network functions (workload) on the open server infrastructure is a similar business motivation. Technically, it requires a high-speed networking interface to VM (or container) guest. DPDK (Data Plane Development Kit) provides such libraries to optimize the I/O performance (with PMD model); the virtualized I/O interface is fragmented because different hypervisor technologies are massively deployed today, such as virtio-pmd for KVM (Kernel-based Virtual Machine), vmxnet3-pmd for VMware, and netvsc-pmd for Hyper-V.

This section has three chapters to cover the virtualization with focus on I/O focus technologies such as SR-IOV and para-virtualization (virtio/vhost). And the changes are introduced with DPDK optimization.

- Chapter 11 focuses on x86 virtualization technology, I/O passthrough, VT-d, and SR-IOV; this helps in the high-performance virtualized (or containerized) workload.
- Chapter 12 introduces KVM-based virtualization interface; discusses virtio and pmd in detail; includes the front-end interface with Linux kernel for cloud workload; and suggests the DPDK-based virtio-pmd for networking function-based workload.

- Chapter 13 focuses on the KVM backend, and vhost pmd design and details. A brief example is added on how to use it.

vmxnet3-pmd and netvsc-pmd are designed for different hypervisors, and the optimization concept is very similar to virtio-pmd/vhost-pmd, so they are not introduced here. VMware vCloud NFV3.0 integrated the DPDK optimization in 2018.

11 Hardware Virtualization

Qian Xu and Rashmin Patel
Intel®

CONTENTS

What is virtualization? Virtualization technology can be traced back to the early 1960s. It only thrives after the x86 CPU enters the multi-core era. Virtualization is the abstract logical resource presented to the workload, and it provides a virtualization layer on physical platform resources such as CPU, memory, and I/O. Virtualization can divide the physical resource into multiple partitions. In essence, virtualization enables multiple workloads to share a set of resources (such as a server). The workload can be migrated to other hardware platform, and as the virtualization is based on software abstraction layer, it decouples the dependency on the underlying hardware platform.

Companies and business organizations have gained significant efficiencies through virtualization because it improved the server system utilization, workload consolidation, dynamic resource allocation, and isolation. Virtualization enables the elastic service model. This is essential for cloud-based IT services. Hardware virtualization (such as Intel® VT) has greatly improved the system performance, which makes the cloud computing from concept to today's hyper-scale cloud use. The telecom operators have gone down the NFV path with a similar model.

I/O virtualization is part of the virtualization technology evolution, and networking function workload has the unique requirement, such as high-speed I/O to the virtual machines (or container instances). High-speed NIC supports the I/O virtualization, and the hardware source can be partitioned into PFs and VFs. DPDK implemented the VF PMD, and it is preferred by VNF tenant.

11.1 X86 VIRTUALIZATION

A server platform consists of three important parts that are to be virtualized: CPU, memory, and I/O. Initially, virtualization is a software implementation, and hypervisor (or VMM; they can be different) provides the abstracted resource for guests.

219

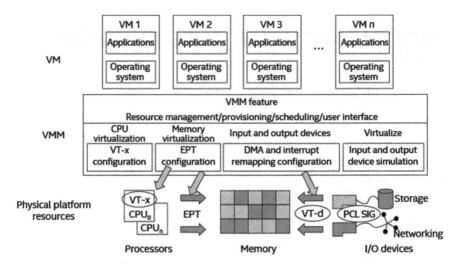

FIGURE 11.1   Hardware-assisted virtualization technology.

To make hypervisor software simple and more performant, the CPU made the design optimization, which helps both performance and security. Intel® architecture provides hardware-assisted virtualization technology (VT-x and VT-d). Figure 11.1 briefly describes it.

11.1.1   CPU VIRTUALIZATION

Intel® VT-x adds the processor extension with the new modes: VMX root mode and VMX non-root mode (as shown below). VMX stands for virtual machine extension. The hypervisor is running on the VMX root mode, while the guest is running on the non-root mode. This can avoid the virtualization vulnerabilities.

In the VMX root mode, all instructions have the same behavior as the legacy architecture. Hypervisor runs in the VMX root mode, and it can execute the privileged instructions at Ring 0.

VM guests will run in the VMX non-root mode. Because multiple guests share the same physical resources, VM guest is designed to run only at Ring 3 for security and isolation. But the guest has its OS, which used to run at Ring 0. In the virtualization architecture, hypervisor is responsible for presenting everything for guest (including its privileged execution with the CPU).

The early instruction virtualization is implemented by the trap-then-emulate approach, and software is the first wave of implementation. The problem is performance challenge and security vulnerability with the privileged instructions. Intel® VT-x introduced a few key innovations on processor:

- VMCS (virtual machine control structure) is primarily for CPU transition from non-root mode to root mode and vice versa. VMCS helps save/restore the virtual CPU status (registers) to accelerate the switching process.
- VMEnter transits into VMX non-root mode

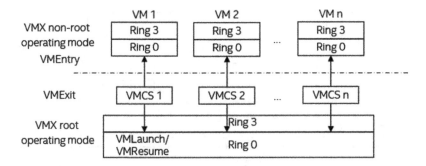

FIGURE 11.2   Intel® VT-x technology.

- VMExit transits into VMX root mode
- VMLaunch/VMResume: VM guest is launched or resumed by hypervisor (Figure 11.2).

11.1.2 MEMORY VIRTUALIZATION

Memory virtualization is important as the guest has its own OS and memory management. It introduces a new layer of address space: the guest physical address. Memory virtualization supports the implementation of the virtual address space by two-level address translation, namely GVA (guest virtual address) -> GPA (guest physical address) -> HPA (host physical address). In this process, the GVA->GPA translation is determined by the guest operating system, usually implemented by the page table, which is referred by CR3, in the guest status area in VMCS. The GPA->HPA translation is decided by the hypervisor. The hypervisor determines this when assigning the physical memory to the guest. The hypervisor will generally use internal data structures to track the address mapping relation.

Prior to virtualization, the x86 architecture supports one-level address translation only, and CR3 refers to the page table for address translation (from the virtual address to the physical address). Memory virtualization adds the two-level address translation; as a result, Intel® VT-x adds an extended page table (EPT) so that two-level address translation (GVA->GPA->HPA) can be assisted by hardware. This greatly reduces the software complexity on memory virtualization, and it improves the virtualization performance significantly.

EPT is described in the below figure. The VM (guest, or client) address translation is given in the following steps:

- The CPU checks the page table and finishes the GPA->HPA translation via EPT. If the address is not hit with EPT TLB lookup, the CPU will check EPT using the traditional page walking mechanism. If there is still no address hit, an exception will be sent, and hypervisor will load the required page into memory to handle this eventually.
- A similar lookup happens from L4 table walk, to L3 walk, to L2 walk, and to L1 walk if required.

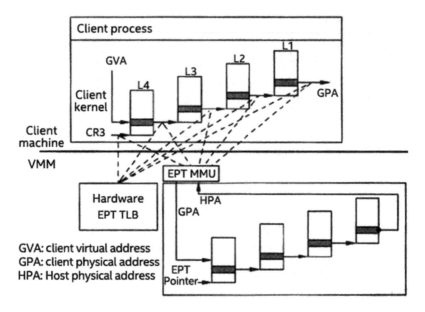

FIGURE 11.3 Extended page table (EPT) concept.

In the worst case, it can be up to 20 memory accesses. Implementing EPT (and its TLB) in the hardware minimizes memory access significantly (Figure 11.3).

11.1.3 I/O VIRTUALIZATION

I/O virtualization is about I/O access between virtual devices and shared physical devices. There are three models for I/O virtualization: I/O full virtualization, I/O paravirtualization, and I/O pass-through. Each is different for guest and host communications (Figure 11.4).

1. I/O Full Virtualization
     This method emulates the actual device in VMM software. All the functions of a device or bus infrastructure (such as device enumeration, identification, interrupt, and DMA) can be emulated in the hypervisor. The virtual machine can only see a setup of virtual platform populated with virtual I/O devices. The hypervisor intercepts guest requests for access to I/O devices and emulates actual hardware through software. This method, transparent to the guest, does not depend on the underlying hardware or modify the operating system. The hypervisor emulates platform I/O seamlessly; hence, the guest operating system is not aware that it is running in an emulated environment, but its efficiency is lower.
2. I/O paravirtualization
     This means that some portion of the guest operating system is aware that it is executing in a virtual machine. I/O virtualization is achieved by a pair

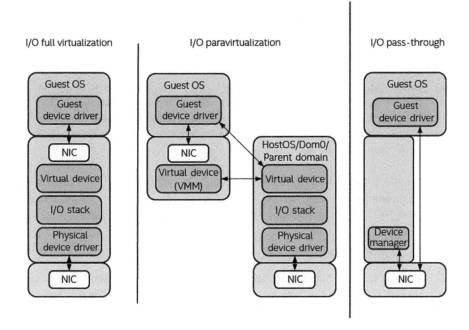

FIGURE 11.4 I/O virtualization options.

of front-end/back-end drivers, which collectively emulate the I/O device operations. The front-end driver is the driver in the guest, and the back-end driver is the driver provided by the hypervisor talking with the guest. The front-end driver sends the guest request to the back-end driver via a special communication mechanism between the guest and the hypervisor. Then, the back-end driver processes the request and sends it to the physical driver. The techniques to achieve paravirtualization may vary on the different hypervisors (KVM, VMware, or Hyper-V). KVM uses the virtio and vhost as the front-end and back-end drivers, respectively, as the main paravirtualization technology, and more details will be discussed in the later chapters. Is I/O paravirtualization better than I/O full virtualization? Both are software approaches; the difference is the communication mechanism between the VM (guest) and VMM (host). For the full virtualization, the emulated I/O device access will cause VMExit. For the paravirtualization option, the negotiation between the front-end and back-end drivers will not cause VMExit in the data transfer, as the back-end driver can access the packet memory, which is allocated by the front-end driver in the guest domain. Although there is the memory address translation from GPA to HPA, the total cost is accepted by public cloud computing infrastructure. Compared to I/O pass-through, the paravirtualization is weak on the performance, but it is based on the software-based virtual device, so it decouples the underlying hardware with the above workload.

3. I/O pass-through is to assign a portion of a physical device (such as a NIC) to the virtual machine. This requires a hardware platform support technology such as Intel® VT-d. The pass-through model can give a near-native I/O performance to a VM, and the drawback is the guest awareness on a specific hardware. When moving the workloads from one machine to another (live migration), the hardware difference will pose a challenge. Paravirtualization is more friendly to live migration, although it also requires that the guest is aware on the virtual device (such as virtio), but most OS is already integrated with virtio support, so this guest awareness is more ready than the pass-through approach.

DPDK supports the I/O virtualization with support on both the paravirtualization and pass-through approaches. With I/O pass-through, DPDK can help VM to achieve the super-high I/O throughput, nearly the physical device performance.

11.2   I/O PASS-THROUGH

I/O pass-through is known for the high I/O performance, close to the native device performance. Intel® VT-d is a big step as it avoids the VMExit for guest I/O access. Today, most PCIe devices are Intel® VT-d-ready. Some known limits on I/O pass-through are given below:

- The x86 platform supports limited numbers of PCIe devices, considering the use and associated cost.
- PCIe pass-through is a challenge to live migration needs. Live migration requires the guest status is saved, while guest workload is moved to another machine and resumed quickly without noticeable difference. The limitation is caused by the hypervisors in the source and destination platforms. The hypervisor on the new platform does not know the internal status of the pass-through device, and therefore cannot restore its status on the destination server.

To address the above challenge, the principles may include the following:

- Only apply the I/O pass-through to a few guests requiring high I/O performance.
- Apply paravirtualization to most guests as software abstraction is important for the cloud principle, such as on-demand software scaling.
- With the hybrid use of I/O pass-through and paravirtualization to the guest, enable them with active/standby by link bonding technique. This is valid for the live migration requirement.
  1. Prior to migration, the pass-through device can be disabled, and guests can swap I/O access to use the software emulation device.
  2. After the guest completes the migration to another host, resume the execution, then insert the pass-through device, swap from the emulated I/O to the pass-through device for high I/O performance.

- NIC can be partitioned with the SR-IOV technology, a NIC can be partitioned into multiple VFs, and then assigning VFs to different guests. Now each guest has the direct VF, this creates the high-performance data path. This sharing method applies to multiple guests, overall this can help to achieve the performance in a scalable approach.

11.2.1 INTEL® VT-D

I/O pass-through is leveraging DMA to realize the high-performing I/O virtualization. The device can support the data transfer to the guest without software overhead. The address in the guest is not the same as the physical address in the host. When a guest asks an I/O device for the data movement, GPA (guest physical address) is sent to the device. The device prefers to use HPA to initiate a DMA, and address translation is needed from GPA to HPA. Intel® VT-d helps the address translation through the DMA remapping functions and direct device assignment.

If the platform is enabled with Intel® VT-d, DMA transfers can be intercepted by the DMA remapping hardware, and the I/O page table will take care of the address translation from GPA to HPA. The mechanism is similar to the CPU page table. The I/O page table supports different page sizes such as 4 KB, 2 MB, and 1 GB. IOTLB is added to speed up the address lookup. IOTLB supports 2 MB and 1 GB HugePages, which has a positive impact on the overall performance.

VT-d introduces the protected domain, and each domain is isolated with the assigned physical memory. The I/O device is partitioned, and only the assigned partition can access the physical memory in this domain. In a virtualized environment, the hypervisor considers each VM as an isolated domain. The restricted access is supported through the DMA remapping hardware unit. More details can be found on "vt-directed-io-spec" at the Intel® Web site (https://software.intel.com/sites/default/files/managed/c5/15/vt-directed-io-spec.pdf) (Figure 11.5).

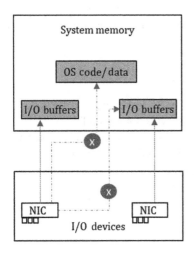
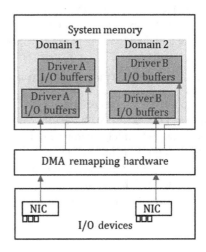

FIGURE 11.5   VT-d: DMA remapping.

Intel® VT-d provides the following capabilities:

- I/O device assignment: Flexible I/O device assignment to VMs and extending the protection and isolation properties of VMs for I/O operations
- DMA remapping: Address translation support for DMA (direct memory access) from devices
- Interrupt remapping: Isolation and routing support for interrupts from devices and external interrupt controllers to appropriate VMs
- Reliability: Recording and reporting of DMA and interrupt errors to system software that may otherwise corrupt memory or impact VM isolation

11.2.2 SR-IOV

The PCIe pass-through mechanism allows a NIC to be passed to a VM directly. This greatly improves the I/O throughput to a VM. In general, NIC is indeed a shared resource, and not every VM can have its own physical NIC and it needs to be shared by multiple VMs. The SR-IOV technique has emerged to respond to this need.

SR-IOV (single root I/O virtualization) is a hardware virtualization specification developed by PCI-SIG. SR-IOV is widely used by the PCIe device, which can be partitioned, and it can support up to 256 virtual functions. This can share NIC among VMs. SR-IOV defines the PF and VF concepts:

- PF (physical function): a PCIe function with SR-IOV extended capability. It is responsible for configuring the device. In theory, a device can support multiple PFs. Today, most NICs only support one PF, and PF cannot be dynamically created and destroyed in the system.
- VF (virtual function): a "lightweight" function and a virtual instance of partial device resource but necessary for basic function. VF can be created and destroyed by PF. Each VF has its own independent PCI configuration space, send and receive queues, interrupts, and other resources. The hypervisor can assign one or more VFs to a VM. Similarly, one or more VFs can be assigned to the container instance (Figure 11.6).

11.3 FREQUENTLY ASKED QUESTIONS

Question: If VT-d is turned on via BIOS on the Intel® server, when passing NIC to a VM, VM fails to start with the error message "IOMMU not found".

Answer: Set "intel_iommu=on" in the Linux boot option, because some Linux kernel versions may not enable CONFIG_INTEL_IOMMU by default.

Question: When trying to run the DPDK program on the host, an error occurs when using the igb_uio module. dmesg says a DMAR error, for example, "dmar: DMAR:[DMA Read] Request device [07:00.1] fault addr 1f732 20000".

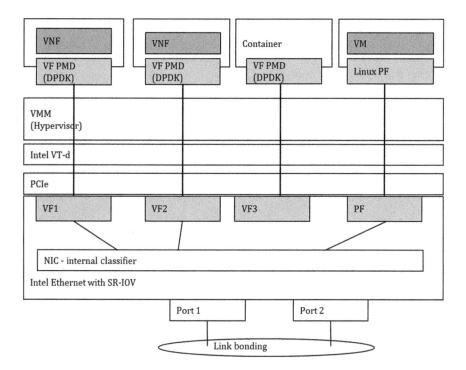

FIGURE 11.6   SR-IOV for VNF.

Answer: There might be two solutions:

- If VT-d pass-through is required (assign PCIe device to a VM), enable "set iommu=pt" in the Linux kernel boot option. This tells the kernel to skip address translation (bypass VT-d). This does not affect the address translation for the VM.
- If it is not necessary to assign a PCIe device to a VM, "disable VT-d" in the BIOS setting, or change the Linux kernel boot option from "intel _ iommu=on" to "intel _ iommu=off".

Question: Similar to the above question, vfio-pci helps the NIC binding to a host (not using the igb_uio module) and then runs the DPDK program. The program seemed to run properly without using "iommu=pt" in the Linux boot option. Why?

Answer: VFIO is an IOMMU user space driver. When a VFIO driver is used, DPDK is responsible for setting the DMA remapping, and it does not need the page table for VM address translation. But for igb_uio, IOMMU address translation cannot be skipped if IOMMU is not configured directly.

12 Virtio

Tiwei Bie
AntFinancial

Changchun Ouyang
Huawei

Heqing Zhu
Intel®

CONTENTS

Virtio is an abstraction for paravirtualized device interface. It is designed for providing hardware-independent, vendor-neutral support for the live migration of tenant workload. This is essential for cloud computing principles. It was developed by Rusty Russell in support of his own virtualization solution called lguest. Later, virtio was widely applied in QEMU (Quick Emulator)-based and KVM (kernel-based virtual machine)-based virtualization technology scenarios. Today, it is widely supported by most guest operating systems, such as Windows and Linux. The front-end driver implemented in the guest operating system is generally called virtio, while the common back-end driver implemented in a hypervisor is called vhost. Compared with the legacy network I/O device (for example, e1000 and RTL8139) emulated simply by software in the hypervisor, virtio, more specifically virtio-net, has a good I/O performance. Earlier, the drawback was that it required the guest to install specific virtio drivers. The guest needed to be aware that it is running in a virtualized environment. Virtio is more than networking (like virtio-net), it also includes the other interface support such as block device. In this chapter, the main focus is on the virtio-net, unless explicitly stated otherwise. Virtio is born and existing long enough in Linux kernel, virtio-net is acceptable as a common practice for public cloud workload, work seamlessly together with virtual switches (such as Open vSwitch), as the network interface to virtual machine (guest or tenant).

Virtio is a highly desired VNF standard interface due to its big success in cloud computing. The abstraction helps to avoid the NFV platform vendor lock-ins. But the kernel-based implementation has one problem. The performance is a challenge in the Linux kernel-based implementation. DPDK introduced the PMD concept for optimizing both the front-end and back-end performance. We will discuss virtio in this chapter and vhost in the next chapter.

12.1   VIRTIO OVERVIEW

There is an extensive use of virtualization technology in cloud computing in which device virtualization plays an important role. Due to the wide range of I/O devices, the features supported by different manufacturers are also different. In general, when the server uses a device, it must first install the device driver, which is closely associated with the device features, and then develop software applications to take advantage of the features of the device, in a virtualized fashion. The process and troubleshooting in the operation and maintenance stages may also be closely associated with the specific device, which is not scalable as open system tends to use multiple device suppliers.

Initially invented as software emulation method virtio has evolved into a standardized device interface, so the mainstream operating systems and applications gradually added the direct support for virtio devices. This is more convenient for the operation and maintenance of the data center.

Compared with the I/O pass-through techniques (such as SR-IOV), virtio is not currently competitive in network throughput, delay, or jitter. So why is still virtio required? A typical problem of I/O pass-through is that the packet received from the physical NIC will be directly delivered to the receive queue of the guest (tenant), or the packet sent by the transmit queue of the guest will be directly delivered to the

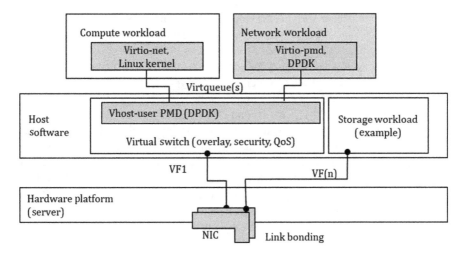

FIGURE 12.1 Typical scenario for virtio.

receive queue of the other guest (VF of the same PF), or the packet is sent directly from the physical NIC bypassing the hypervisor. But in many scenarios, the network packets must first be processed by the host (such as firewalls and load balancing, virtual switch) before being passed to the guest. I/O pass-through does not support the live migration of VMs with multiple hardware vendors; VF interface is vendor specific. The current NIC also lacks adequate flexibility in traffic classification capability for cloud computing. Notably, some industry players have started to implement virtio interface in NIC. Hardware-assisted virtio will fill the performance gap we mentioned above.

Figure 12.1 shows a typical scenario with virtio devices used in a data center server. The server connects the physical NIC(s) to a VM's workloads using a virtual switch. There is a DPDK vhost-user PMD in the virtual switch to realize the virtio back-end network device driver. The VM has the virtio as the front-end network device driver. It has two options: Running the virtio-net kernel module for compute-intensive workload or running the DPDK-enabled virtio-pmd for network-intensive workload, which provides a higher networking throughput.

The front-end and the back-end exchange data with each other through virtual queue(s) of the virtio. In this way, the network data from the VM are sent to the virtual switch, then pass through switching process, and finally enter the external network via a physical NIC (Figure 12.1).

12.2 VIRTIO SPECIFICATION

Virtio has three specification versions: 0.95, 1.0, and 1.1, defined and evolved in community under multiple years of effort. The specified interfaces include PCI (peripheral component interconnect), MMIO (memory-mapped I/O), and channel I/O, which is newly added in Spec 1.0. DPDK focuses on packet processing, which currently supports virtio-net in PCI interface mode.

Spec 1.0 is compatible with the previous Spec 0.95. It refers to the modes defined earlier as "legacy" and the modes later defined as "modern". Legacy PCI is currently the most widely used mode. Modern PCI known as PCIe is supported in Linux kernel 4.0 or above. Modern and legacy PCI devices are different in terms of parameters and usage modes, but in Linux kernel 4.0, they are implemented using the same virtio device driver. The driver will automatically be selected, it depends on the PCI device emulated using QEMU, and it can be either legacy or modern.

Spec 1.1 is a very new effort, and it is also designed with backward compatibility with the previous Specs 0.95 and 1.0. In virtio1.1, it refers to the virtqueue layout defined in earlier specs as "split virtqueue" and defines a new virtqueue layout named "packed virtqueue". The packed virtqueue is supported in Linux kernel 5.0 or above. Similar to the case of "legacy" device and "modern" device, the "split virtqueue" and "packed virtqueue" are implemented using the same virtio device driver, and the driver will handle if the device supports "packed virtqueue" or not.

Virtio defines a virtqueue (virtual queue) interface on top of the structure of PCI (transport layer), which conceptually connects the front-end driver to the back-end driver. The driver may use one or multiple queues, and the exact number of queues depends on the use case. For example, the virtio network driver uses two virtual queues (one for receiving and one for transmitting), while the virtio block driver uses one virtual queue.

When virtio is implemented using the PCI interface for networking, the hypervisor will use the back-end driver to emulate a PCI device and this device can be added into VM launch configuration or hot-plugged into VM at runtime. Next, we will take the widely used legacy PCI device as an example to explain in detail the overall framework and principles of virtio. Some parts also involve the modern mode. For brevity, unless otherwise specified, the driver refers to the virtio front-end driver and the device refers to the PCI device emulated by the virtio back-end driver (such as vhost) in the following.

12.2.1 DEVICE INITIALIZATION

Device initialization has the following five steps. When the initialization is successful, the device can be used in the guest.

- After the device is rebooted (or powered on), the system discovers the device.
- If the guest operating system sets the device status as Acknowledge, the device is identified.
- If it sets the device status as Driver, it finds an appropriate driver.
- Installation and configuration of the device driver has the following steps: negotiation using feature bits at front-end and back-end, initialization of the virtual queue(s), optional MSI-X installation and device-specific configuration, etc.
- Then the status of the device is set as Driver_OK. If there is an error in the process, then the status is set as Failed.

1. Device Discovery

Spec 1.1 defines/reserves 24 types of virtio devices, as shown in Table 12.1. (Some device IDs are reserved for devices which are not currently defined.) Here, virtio device ID represents the different kinds of devices; it is beyond the network interface. As said, virtio-net is the DPDK initial focus. Most NICs are PCIe devices and use the PCI scheme; the virtio device ID is not necessarily the same as the PCI device ID or the device IDs of other specific implementation schemes.

The virtio PCI device follows the PCI standard specification and uses the PCI standard configuration space and I/O area. The PCI vendor ID of the virtio device is 0x1AF4. The PCI device ID ranges from 0x1000 to 0x107F. 0x1000–0x103F are used for legacy (virtio0.95) devices, while 0x1040–0x107F are used for modern (virtio1.0) devices. For example, PCI device ID 0x1000 represents a legacy virtio NIC, while 0x1041 is a modern virtio NIC. Both correspond to a NIC device whose virtio device ID is 1 (Table 12.1).

TABLE 12.1
Virtio Device ID

Virtio Device ID	Virtio Device
0	Reserved (invalid)
1	Network card (also known as NIC)
2	Block device
3	Console
4	Entropy source
5	Memory ballooning (traditional)
6	ioMemory
7	Rpmsg
8	SCSI host
9	9P transport
10	Mac80211 WLAN
11	Rproc serial
12	Virtio CAIF
13	Memory balloon
16	GPU device
17	Timer/clock device
18	Input device
19	Socket device
20	Crypto-device
21	Signal distribution module
22	pstore device
23	IOMMU device
24	Memory device

2. Configuration Space of Legacy Virtio Devices

The legacy mode uses PCI BAR0 for device configuration. The configuration parameters are shown in Table 12.2.

If a legacy device is configured with an MSI-X (message signaled interrupt-extended), additional bits are added to support the read/write domains, as illustrated in Table 12.3.

Following these common virtio parameters, there may be some device-specific (e.g., NIC) configuration parameters, as shown in Table 12.4.

3. Configuration Space of Modern Virtio Devices

Unlike the legacy device which uses BAR0, the modern device is enhanced with the new PCIe configuration mechanism (for example, which BAR to be used and the offset address starting from the BAR space). Spec 1.0 defines four kinds of configuration information: common configuration, notifications, ISR status, and device-specific configuration.

Figure 12.2 shows a common configuration in the modern device configuration space. Compared with the legacy device in Table 12.2, the new configuration enhances feature negotiation, queue size, and so on. It also includes the MSI-X additional configuration for legacy devices in Table 12.3 (Figure 12.2).

TABLE 12.2

Legacy Virtio Devices: Common Configuration Space

Bits	32	32	32	16	16	16	8	8
Read / write	R	R+W	R+W	R	R+W	R+W	R+W	R
Purpose	Device features— bits 0:31	Driver features— bits 0:31	Queue address	*queue_size*	*queue_select*	Queue notify	Device status	ISR status

TABLE 12.3

Legacy Virtio Devices: MSI-X Additional Configuration Space

Bits	16	16
Read/write	RW	R+W
Purpose (MSI-X)	*Config_msix_vector*	*Queue_msix_vector*

TABLE 12.4

Legacy Virtio Devices: Device-specific Configuration Space

Bits	Device specific	···
Read/write	Device specific	
Purpose	Device specific	

```
struct virtio_pci_common_cfg {

        /* About the whole device. */

        uint32_t device_feature_select;   /* read-write */

        uint32_t device_feature;          /* read-only */

        uint32_t guest_feature_select;    /* read-write */

        uint32_t guest_feature;           /* read-write */

        uint16_t msix_config;             /* read-write */

        uint16_t num_queues;              /* read-only */

        uint8_t device_status;            /* read-write */

        uint8_t config_generation;        /* read-only */


        /* About a specific virtqueue. */

        uint16_t queue_select;            /* read-write */

        uint16_t queue_size;              /* read-write, power of 2. */

        uint16_t queue_msix_vector;       /* read-write */

        uint16_t queue_enable;            /* read-write */

        uint16_t queue_notify_off;        /* read-only */

        uint32_t queue_desc_lo;           /* read-write */

        uint32_t queue_desc_hi;           /* read-write */

        uint32_t queue_avail_lo;          /* read-write */

        uint32_t queue_avail_hi;          /* read-write */

        uint32_t queue_used_lo;           /* read-write */

        uint32_t queue_used_hi;           /* read-write */
};
```

FIGURE 12.2   Modern virtio device: common configuration space.

In the following, we will explain several key parameters in detail:

1. Device Status
     When the virtio driver initializes a virtio device, the initialization prog-
ress can be reflected through the device status. Below are the five statuses
defined in legacy devices:
     0: When the driver writes "0", it indicates that the device needs a restart.
     1: "Acknowledge" indicates that the guest operating system has found a
        valid virtio device.

2: "Driver" indicates that the guest operating system has found an appropriate driver (like the virtio NIC driver).

4: "Driver_OK" indicates that the driver has been installed successfully and
the device is ready for use.

128: "FAILED" indicates that an error occurs in the driver installation
process.

The modern device includes two more statuses:

8: "FEATURES_OK" indicates that the driver and the device features have
negotiated successfully.

64: "DEVICE_NEEDS_RESET" indicates that the device has failed and
needs a restart.

2. Feature Bits

Devices (from back-end) and drivers (in front-end) have their respective feature bits. The feature bits of the modern device support 64 bits, and
those of the legacy device support 32 bits only. Both devices and drivers can
provide self-supported feature sets through feature bits. In the initialization
process, the driver reads the feature bits of the device and then selects those
that can be supported on their own as a driver. This completes the feature
negotiation between the driver and the device. The specific feature bits are
listed as follows:

- 0–23: Feature bits of a specific device. Each device has its own feature
definition. An example is that the virtio-net device supports 24 features,
of which VIRTIO _ NET _ F _ CSUM uses bit 0 to indicate whether it
supports checksum offload on the host side and VIRTIO _ NET _ F _
GUEST _ CSUM uses bit 1 to indicate whether it supports checksum
offload on the guest side. Most NICs support checksum offload, and the
guest can defer the checksum calculation to the host NIC, if the packet
destination is out of this current system.
- 24–38: Reserved for the extension of queue and feature negotiation
mechanism. For example, VIRTIO _ F _ RING _ INDIRECT _
DESC uses bit 28 to indicate whether the driver supports the indirect
descriptor table, which is necessary to carry out larger data transfers.
- 39–: Reserved for future expansion (only applicable for modern devices).

3. Interrupt Configuration

Both modern and legacy devices support two interrupt sources (device
interrupt and queue interrupt) and two interrupt modes (INTx and MSI-
X). Each device has only one device interrupt source, while each queue
has one queue interrupt source. However, the exact number of interrupts
also depends on the interrupt mode. In INTx mode, one device supports
one interrupt only, so the device interrupt source and the queue interrupt
source have to share this interrupt. MSI-X supports multiple interrupts.
Each individual interrupt is called an interrupt vector. Supposing that there
are n queues, there will be n queue interrupts plus one device interrupt in
the device, so there are n+1 interrupts in total. The n+1 interrupts can be

flexibly configured. Any one of the interrupt sources can configure and use any one of the interrupt vectors.

INTx is rare now, and new systems generally support a more powerful MSI-X mode. Next, we will introduce MSI-X-related settings. After the MSI-X interrupt is enabled, the legacy device can use the two registers in MSI-X additional configuration, as shown in Table 11.3, to map the device and queue interrupts into the corresponding MSI-X interrupt vectors (corresponding to the configuration vector and queue vector). These two registers are of 16 bits and are readable and writable. Interrupts are mapped by writing a valid interrupt vector value (the range of valid value: 0x0–0x7FF). After having an interrupt, the device or queue will notify the driver through this interrupt vector. Writing VIRTIO _ MSI _ NO _ VECTOR (0xFFFF) will close the interrupt and cancel mapping. Reading these two registers will return the mapping to the interrupt vector of the specified interrupt source. If there is no mapping, return VIRTIO _ MSI _ NO _ VECTOR. In the modern device, these two registers are directly included in the common configuration. Their usage is similar to that of the legacy device. Resources should be allocated when an interrupt source is mapped onto the interrupt vector. If the mapping fails, reading the register value returns VIRTIO _ MSI _ NO _ VECTOR. After successful mapping, the driver must read the values of these registers to confirm successful mapping. If the mapping fails, you can try to map fewer interrupt vectors or close MSI-X interrupts.

4. Device-specific Configuration

This configuration space contains device-specific information. Taking a NIC device as an example, the legacy device can define MAC address and status information and the modern device can add more information such as the maximum number of queues. The mechanism provides the feature extension and flexibility.

12.2.2 VIRTQUEUE

Virtqueue is the actual data structure connecting the virtio front-end driver in the guest operating system and the back-end driver in the host. Its schematic diagram is shown in Figure 12.3 (split virtqueue) and Figure 12.4 (packed virtqueue), and the actual scheme is decided by one Boolean bit (packed_ring) at the virtqueue creation phase, which is a result of front-end and back-end negotiation.

12.2.2.1 Virtqueue Initialization

In general, virtqueue initialization is a step after the device initialization. The virtqueues can be initialized in below steps:

- Write the virtqueue index to the queue _ select register.
- Read the queue _ size register to get the size of the virtqueue. If the queue size is 0, the queue is not available. (The queue size of the legacy

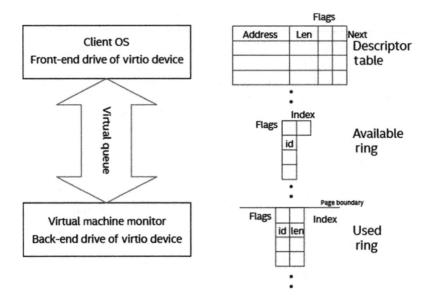

FIGURE 12.3   Virtio split virtqueue diagram.

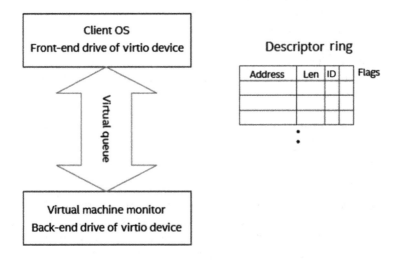

FIGURE 12.4   Virtio packed virtqueue schematic diagram.

device must be specified by the device, while in the modern device, the driver may write a smaller value into the queue _ size register to reduce the use of the memory.)

- Assign the guest memory to the queue, and write the guest physical address into the queue _ desc, queue _ driver, and queue _ device registers.

- If the MSI-X interrupt mechanism is enabled, select an interrupt vector for the virtqueue, write the MSI-X vector into the `queue _ msix _ vector` register, and then read the field again to confirm that the correct value is returned.

12.2.2.2   Split Virtqueue (Virtio1.0)

The virtqueue consists of a descriptor table, available ring, and used ring. The descriptor table refers to the actual data blocks to be transmitted, and both the rings point to the descriptor table. The available ring is for the guest to update which descriptor to use, and the used ring is returned by the host after its use (either read or write). The actual data exchange, via virtqueue, is between the front-end and back-end.

12.2.2.3   Descriptor Table

Each descriptor element in the table represents a data buffer of the guest VM, and the data buffer is used for transmitting data between the guest and the hypervisor (host). If the data being transmitted between the guest and the hypervisor exceed the size limit, then multiple descriptors can form a descriptor chain (indirect descriptor) to carry the large data. Each descriptor, as shown in Figure 12.3, has four attributes:

- *Address*: Guest physical address of the data buffer.
- *Len*: Length of the data buffer.
- *Next*: Address of the next descriptor in the chain.
- *Flags*: Flags representing certain attributes of the current descriptor, for example whether Next is valid or not (if invalid, the current descriptor is the end of the entire descriptor chain) and whether the current descriptor is writable to the device.

12.2.2.4   Available Ring

The available ring points to the descriptor. It is provided (written) by the guest driver and used (read) by the hypervisor device. If the host takes the descriptor in the available ring, the data buffer may be either writable or readable. The writable buffer is provided by the driver and written into the device used for sending the host data to the guest. The readable buffer is used for sending the guest data to the host. Each entry of the available ring has three attributes, as shown in Figure 12.3,

- *Ring*: An array for storing the descriptor pointers (id).
- *Index*: Location of the next available descriptor written by the driver (guest).
- *Flags*: Flags representing certain attributes of the available ring, for example whether it is necessary for the device to send an interrupt to the driver after using the entry of the available ring.

12.2.2.5   Used Ring

The used ring points to the descriptor. It is provided (written) by the hypervisor device and used (read) by the guest driver. After the host completes the use of the descriptor taken from the available ring, the host will return the descriptor into the used ring so that the guest driver can take it back for another use. The entry of the used ring, as shown in Figure 12.3, has three attributes:

- Ring: An array for storing the used elements. Every used element includes a descriptor pointer (id) and a data length (len).
- Index: Location of the next used element written by the device.
- Flags: Flags representing certain attributes of the used ring, for example whether it is necessary for the driver to send a notification to the device after recycling the entries of the used ring.

12.2.2.6 Packed Virtqueue (virtio1.1)

Packed virtqueue is the latest design, and it mainly consists of the following:

- A descriptor ring,
- A driver event suppression structure (guest),
- A device event suppression structure (host).

It is an alternative virtqueue layout to split virtqueue, and it is more compact. The descriptor table, available ring, and used ring of split virtqueue are squashed into one descriptor ring in packed virtqueue. The use of packed virtqueue is negotiated by the VIRTIO _ F _ RING _ PACKED feature bit. Each descriptor's flags (AVAIL and USED flags) indicate whether the processing has been completed by the front-end or back-end drivers. The primary motivation behind the implementation of the packed virtqueue layout is to be hardware-friendly; for example, the new layout will allow the hardware (such as FPGA) to fetch multiple descriptors using one PCIe bus transaction.

12.2.2.7 Descriptor Ring

Each descriptor element in the descriptor ring represents a data buffer of the guest VM, and it supports the data exchange between the guest and the hypervisor. Multiple continuous descriptors can be chained to describe a scatter buffer list. Every descriptor has the following four attributes:

- *Address*: Guest physical address of the data buffer.
- *Len*: Length of the data buffer.
- *ID*: Buffer ID managed by the guest driver (changed).
- *Flags*: Flags representing certain attributes of the current descriptor, for example whether the next descriptor is chained with the current descriptor or not and whether the descriptor available to the device.

Both of the front-end driver and back-end driver are expected to internally maintain a single-bit ring wrap counter which should be initialized to 1. The counter maintained by the front-end driver is called the Driver Ring Wrap Counter. And the counter maintained by the back-end driver is called the Device Ring Wrap Counter.

12.2.2.8 Event Suppression

The driver event suppression structure is read-only to the device, and it includes the information for reducing the number of driver events generated by the device. The device event suppression structure is write-only to the device, and it includes the information for reducing the number of device events generated by driver. The event suppression structure consists of the below fields:

- *Descriptor event offset*: The event offset of the descriptor ring change event.
- *Descriptor event wrap*: The wrap counter of the descriptor ring change event.
- *Descriptor event flags*: The flags of the descriptor ring change event, which include enable, disable, and desc (which is valid only if VIRTIO_F_RING_EVENT_IDX is negotiated).

12.2.3 DEVICE USAGE

There are two steps to use the device: The driver provides the device with the data buffer through the descriptor table and the available ring; the device then sends the descriptor to the driver through the used ring after using that descriptor. For example, the virtio network device has two virtqueues: a send queue (TXQ) and a receive queue (RXQ). The driver sends the packets to TXQ (read-only for the device) and then releases the packets after being sent by the device. While receiving a packet, the device writes the packet to RXQ and the driver receives and processes the packet in the used ring.

12.2.3.1 Split Virtqueue

1. The guest provides the memory buffer to the device.

    In the guest operating system, the driver provides the device with the memory buffer in the following steps:
    - Assign the information such as the address and length of the data buffer to an idle descriptor.
    - Add the descriptor pointer to the header of the available ring of the virtqueue.
    - Update the header pointer of the available ring.
    - Write the virtqueue identification number to the queue notify register. This is to notify the device.

2. The host uses and returns the data buffer.

    After using the data buffer (the operation may be reads, writes, or partial reads or writes, depending on the type of device), the device fills the used ring with the used buffer descriptor and notifies the driver by the interrupt in the following steps:
    - Add the header pointer of the used buffer descriptor to the used ring.
    - Update the header pointer of the used ring.

- Notify the driver using interrupt, and the interrupt mode is based on MSI-X on/off.
- IN_ORDER feature support.

When VIRTIO _ F _ IN _ ORDER feature is negotiated, the driver will use the descriptors in ring order: starting from offset 0 in the descriptor table and wrapping around at the end of the table. And the device must use the descriptors in the same order in which they have been made available. The device can just write out a single used ring entry with the ID of the header entry of the descriptor chain of the last buffer in a batch to notify the use of the batch of buffers to the driver.

12.2.3.2   Packed Virtqueue

1. The guest provides the memory buffer to the device.
   The guest driver provides the memory buffer in the following steps:
   - Assign the physical address and length of the data buffer to the next descriptor in the descriptor ring.
   - Update descriptor's AVAIL/USED bits based on the value of Driver Ring Wrap Counter to make it available to the device. And change the value of the Driver Ring Wrap Counter if the last descriptor in the ring was just made available.
   - Write the virtqueue identification number to the queue notify register. This is to notify the device.
2. The host uses and returns the data buffer.
   After using the data buffer (the operation may be reads, writes, or partial reads or writes, depending on the type of device), the host returns the used descriptor with the buffer ID and notifies the guest driver by the interrupt. The detailed process is as follows:
   - Write a used descriptor with the information such as buffer ID, flags, and data length (if needed).
   - Update descriptor's AVAIL/USED bits based on the value of Device Ring Wrap Counter to make it available to the driver. And change the value of the Device Ring Wrap Counter if the last descriptor in the ring was just used.
   - Notify the driver using interrupt, on the basis of MSI-X on/off.
3. IN_ORDER feature support
   When VIRTIO _ F _ IN _ ORDER feature is negotiated, the device must use the descriptors in the same order in which they have been made available. And the device can just write out a single used descriptor with the buffer ID of the last descriptor in a batch to notify the use of the batch of buffers to the driver.

12.3   VIRTIO NETWORK DEVICE DRIVER

The virtio specification supports many device types, and the network device might be the most complex device. The Linux kernel driver (virtio-net) has comprehensive features widely used for cloud computing workloads in guests, where the packet throughput requirement is not high, that is typically less than 800kpps.

While the DPDK driver (virtio-net-pmd) offers a higher performance, it uses the poll mode with one or multiple cores (for multi-queues). This is mainly used for VNF/NFV deployment, network-intensive workload in cloud environment.

12.3.1 VIRTIO-NET

The Linux kernel driver for the virtio network device, virtio-net kernel module, consists of three layers: an underlying PCIe device layer, a virtqueue layer in the middle, and a network device layer in the top. Here is an example on Linux kernel v4.1.0 to illustrate the three layers and relationships.

1. The PCIe Device Layer

   The underlying PCIe device layer is responsible for detecting the PCIe device and initializing the corresponding device driver. As shown in the module composition schematic diagram in Figure 12.5, the source file is implemented with the C language.

   virtio _ driver and virtio _ device are the abstract classes of virtio drivers and devices, in which some public properties and methods,

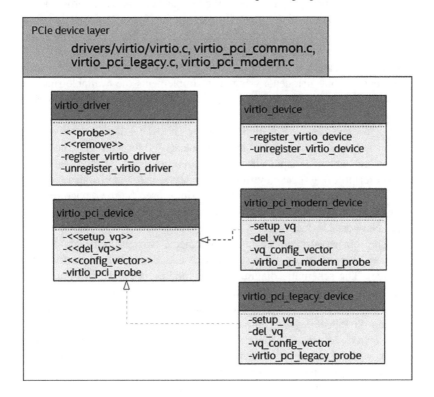

FIGURE 12.5   PCIe device layer.

such as registration with the kernel, necessary for all virtio devices are encapsulated.

virtio _ pci _ device represents an abstract virtio PCIe device. virtio _ pci _ probe is a callback function registered with the Linux kernel system. The kernel system will call this function for further processing after it discovers a virtio device. setup _ vq, del _ vq, and config _ vector are the corresponding function interfaces. A concrete implementation (virtio _ pci _ modern _ device or virtio _ pci _ legacy _ device) is used to set up a virtqueue, delete a virtqueue, and configure an interrupt vector.

virtio _ pci _ device has two concrete implementations: One is virtio _ pci _ modern _ device for implementing the modern protocol, and the other is virtio _ pci _ legacy _ device for implementing the legacy protocol. The two implementations have their respective probe functions: virtio _ pci _ legacy _ probe and virtio _ pci _ modern _ probe. A successful device probe will create the corresponding virtio _ pci _ device.

setup _ vq is responsible for creating vring _ virtqueue, which is the virtqueue layer in the middle.

2. The Virtqueue Layer

The virtqueue layer implements the virtqueue in the virtio protocol. The schematic diagram of the module is shown in Figure 12.6. The top-layer vring_virtqueue structure represents a virtqueue, where vring is the associated data structure and virtqueue connects the device and implements queue operations. In the data structure of vring, vring_desc implements the descriptor table in split virtqueue, vring_avail implements the available ring, and vring_used implements the used ring. In the data structure of vring_packed, vring_packed_desc implements the descriptor ring in packed virtqueue and vring_packed_desc_event implements the driver and device event suppression structures. virtqueue_add in virtqueue is used for adding the descriptor to the available ring (split virtqueue) or making the descriptor available in the descriptor ring (packed virtqueue) for device use, and virtqueue_get_buf in virtqueue is used for getting the descriptor used by the device from the used ring (split virtqueue) or descriptor ring (packed virtqueue).

3. The Network Device Layer

The top network device layer implements two abstract classes: virtio devices (virtio_net_driver::virtio_driver) and network devices (dev::net_device). The schematic diagram of the module is shown in Figure 12.6. virtio_net_driver is a concrete implementation of the abstract virtio device for the network device, which uses the underlying PCIe device and the middle virtqueue layer to implement the packet sending and receiving of the network device and other control features. Dev is a concrete implementation of the Linux abstract network device, which implements the Linux net_device_ops interface through virtnet_netdev and the Linux ethtool_ops interface through virtnet_ethtool_ops, so that the Linux system can operate this virtio network device in the same way as it operates a common NIC.

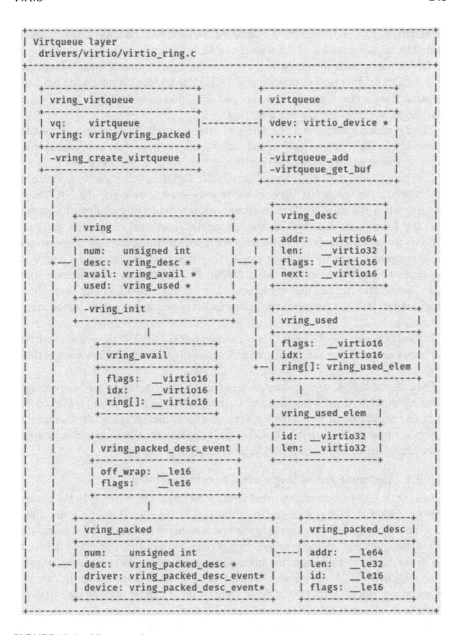

FIGURE 12.6 Virtqueue layer.

12.3.2 VIRTIO-NET-PMD

DPDK user space-based virtio network driver implements the same virtio PCIe protocol as the Linux kernel driver. An example is introduced by DPDK19.02. Currently, the DPDK driver supports both legacy and modern devices. The main implementation is under the directory drivers/net/virtio/ and consists of three layers:

an underlying PCIe device layer, a middle virtqueue layer, and a top network device layer. The implementation of the underlying PCIe device layer is more common in the DPDK public component. virtio_pci.c and virtio_pci.h mainly include tool functions such as reading the configuration in PCIe. The middle virtqueue has basically the same vring and vring_desc structure and definition as in the Linux kernel driver in virtqueue.c, virtqueue.h, and virtio_ring.h.

The top network device layer implements various interfaces of rte_eth_dev, which are mainly in files virtio_ethdev.c and virtio_rxtx.c. Virtio_rxtx is responsible for receiving and sending packets, while virtio_ethdev is for setting devices.

Compared with the Linux kernel driver, the DPDK user space network driver only supports the virtio NIC device, so we only need to consider the NIC device scenario when it comes to the overall architecture and optimization. To be specific, the Linux kernel driver implements public functional component for basic virtio on which some feature negotiations, virtqueue settings, etc., are implemented. These components can be shared among all the virtio devices such as virtio network devices and virtio block devices. But in DPDK, in order to improve efficiency, some basic features are merged and implemented in the top network device layer. Next, we will detail DPDK's performance optimization of single-frame and jumbo-frame processing.

In most other places, the basic process and features of the DPDK user space driver and Linux kernel driver are consistent and compatible, so we will not discuss these here.

Overall, the DPDK user space driver makes full use of the DPDK's advantages in the architecture (SIMD instructions, hugepages mechanism, round robin mechanism, and avoidance of switching between users and memory) and the feature that only NIC optimization is needed. Although it implements the same virtio protocol as the kernel driver, the overall performance is greatly improved.

12.3.2.1   Optimization of Single mbuf Packet Receiving

A packet that can be put into an mbuf struct is called a single mbuf. In the general network packet receiving process, the front-end driver should assign a descriptor from the idle vring descriptor table to each packet, fill in the guest buffer-associated information, and update the available ring entries and avail idx, and then the back-end driver reads and operates the updated available ring. For virtio implementations in QEMU/KVM, the number of vring descriptors is generally set to 256. For the receiving process, HEADROOM in front of mbuf can be used as the virtio net header space, so one packet needs one descriptor only.

Here comes a typical performance problem. Since the front-end driver and the back-end driver generally run on different CPU cores, the update of the front-end driver and the read of the back-end driver may trigger cache migration of avail ring entries between cores, which is time-consuming. In order to solve this problem, DPDK offers an optimized design, as shown in Figure 12.7, fixing the mapping between the avail ring entry and the descriptor table entry. That is to say, all the head_idx entries of the avail ring point to a fixed location of vring descriptors. (For the receiving process, the fixed mapping of the avail ring is 0-> descriptor table 0, 1->1,..., 255->255.) The update of the avail ring needs to update only the pointer of

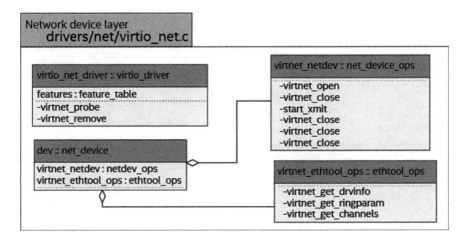

FIGURE 12.7 Network device layer.

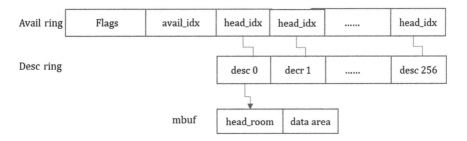

FIGURE 12.8 Optimization of fixed available rings.

the ring itself. A fixed avail ring can avoid cache migration between different cores, eliminate the allocation and deallocation of vring descriptors, and facilitate further acceleration using SIMD instructions.

It should be stressed that the optimization is only for receiving single mbuf for split virtqueue. For packed virtqueue, the avail ring, used ring, and desc table are already squashed into one single descriptor ring by design.

12.3.2.2   Support for Indirect Feature in Packet Sending

Indirect feature refers to that virtio front-end and back-end drivers support VIRTIO_F_RING_INDIRECT_DESC flags after negotiation, indicating that the drivers support the indirect descriptor table. As described earlier, sending a packet requires at least two descriptors. With the support for indirect feature, any packet to be sent, either single frame or jumbo frame (see Section 6.6.1 for the introduction of jumbo frame), shall need one descriptor only. The descriptor points to a memory area of the indirect descriptor table additionally assigned by the driver. Every descriptor in the indirect descriptor table points to a virtio net header and a data area of each mbuf (one mbuf for a single frame; multiple mbufs for a jumbo frame). As shown in Figure 12.8, every descriptor in the virtio queue descriptor table points to a memory

Vring desc table

L: len; F: flag; N: next

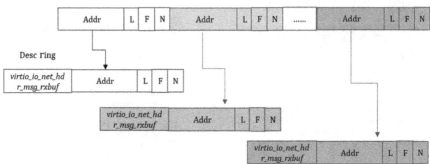

FIGURE 12.9   Indirect descriptor table.

area of the indirect descriptor table. The descriptors (eight assigned to DPDK at present) in this area are connected to form a chain by next domains. The first descriptor is used to point to the virtio net header, while the remaining seven descriptors can point to up to seven data areas of a jumbo frame (Figure 12.9).

This optimization targets packet sending, making all the sending packets use one descriptor only. Tests verify that it can indeed improve performance.

12.3.2.3   IN_ORDER Feature

IN_ORDER feature refers to that the virtio driver and device support VIRTIO _ F _ IN _ ORDER flag after negotiations, indicating that the driver will use the descriptors in the ring order and the device will use the descriptors in the same order in which they have been made available. This optimization targets both sending and receiving of the packets. It will simplify the allocation and deallocation of vring descriptors and make the vring descriptors usage more predictable. All vring descriptors will be linked at the initialization time. And the descriptors will be used sequentially during packet sending and receiving. It also allows further acceleration using SIMD instructions.

12.3.2.4   Multi-queue

Multi-queue support allows I/O-intensive workload to use more cores for packet in/out on the same device at the same time with no contention. This is an important feature to leverage multi-core CPUs' parallel processing capabilities. The device queues which allow the driver to receive packets are RXQs, and the device queues which allow the driver to send packets are TXQs. When this feature is enabled, typically, different CPU cores will manage different sets of RXQs and TXQs, respectively. The number of RXQs and TXQs will be set by the application via DPDK standard APIs during the device setup. And DPDK's RX and TX APIs are queue aware, so the application can easily use different CPU cores to send/receive packets via different TXQs/RXQs on the same device at the same time with no contention.

To use multi-queue, the VM needs to be launched with multi-queue enabled:

```
qemu-system-x86_64 … -chardev socket, id=char0,path=/tmp/
vhost-user \
-netdev type=vhost-user, id=net0,chardev=char0,vhostforce,
queues=$QUEUE_NUM \
-device virtio-net-pci, netdev=net0,mq=on, vectors=$((2 *
$QUEUE_NUM + 2))
```

Normally, the number of vectors should be 2 * queue_num + 2 to allow the guest to assign different interrupt vectors to each TX queue/RX queue, control queue, and configuration change.

12.4 VIRTIO-USER FOR CONTAINER

Container is very popular because of low overhead, fast boot-up time, etc. The virtual device in DPDK, virtio-user, with unmodified vhost back-end, is designed for high-performance container networking.

Virtio-user is a combination of the virtio driver and vhost front-end. In virtio-user, the driver's setups done on the device will be translated into vhost commands/messages and sent to the vhost back-end. It will also share the process's DPDK memory to the back-end with the standard vhost mechanisms. So, it allows a DPDK process to connect to a vhost-user-based virtual switch (e.g., OVS-DPDK) directly through the standard vhost-user interface (i.e., a UNIX socket path).

The overview of using virtio-user to accelerate the container networking is shown in Figure 12.10 (obtained from DPDK how-to [2]) (Figure 12.10):

The whole DPDK application running in the container will connect to a vhost-user port of the vSwitch or vRouter via the standard UNIX socket-based interface.

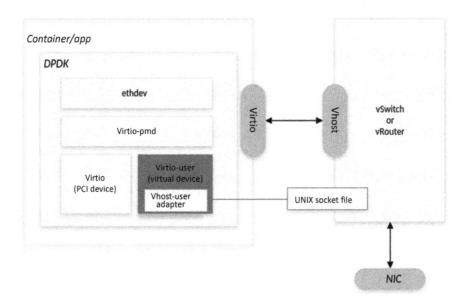

FIGURE 12.10 Virtio for container networking.

In case of VM, QEMU always shares the whole guest physical memory of the VM to the vhost back-end, which is not feasible for container. So, in virtio-user, only the memory managed by DPDK will be shared to vhost back-end for packets TX/RX.

12.5 CONCLUSION

In this chapter, we first briefly described the typical scenario of paravirtualization virtio and then discussed virtio techniques in detail, including configuration at the device level and virtqueue level, the device usage. The third section introduced two different front-end driver designs for the virtio network devices, including Linux kernel and DPDK user space drivers, DPDK optimization technology, and virtio-user. We will discuss the back-end driver technique for virtio devices—vhost—in the next chapter.

REFERENCE

1. http://doc.dpdk.org/guides/howto/virtio_user_for_container_networking.html.

13 Vhost-User

Tiwei Bie
AntFinancial

Heqing Zhu
Intel®

CONTENTS

The previous chapter introduced the front-end driver design of virtio-net device. In this chapter, the corresponding back-end driver is illustrated. Its evolution goes through multiple stages below performance pushes the technology evolution.

1. virtio-net back-end driver.
2. vhost-net.
3. vhost-user (by DPDK).
4. vhost acceleration by hardware.

13.1 VHOST EVOLUTION

13.1.1 INITIAL BACK-END

From the very beginning, the virtual queues, message notifications, and interrupts are essential for virtio-net back-end drivers. The virtual queue mechanism supports the packet/data exchange between the guest and the host. The message is the notification between the guest and the host. The interrupt mechanism is used as well between the host and the guest.

Figure 13.1 shows the system architecture diagram of the early back-end module. Here, KVM is a kernel virtualization of x86 hardware; QEMU depends on KVM to emulate the full emulator environment such as CPU and I/O device; TAP is a virtual network device in the Linux kernel and it is a known interface between the Linux kernel and user space program, which provides interface to QEMU user space emulator (Figure 13.1).

When the guest sends a packet, it uses the message (highlighted as path 2) to notify KVM and goes back to user space QEMU process. Next, QEMU begins to read and write the TAP device (datapath, highlighted as path 1). This architectural model requires the processing among the host, the guest, and QEMU, which results in multiple times of data copying and CPU context switching. As a result, the performance is not high and it is limited by the datapath and the message path.

- In the datapath, two types of packet copying are seen, one from the TAP device to QEMU and the other from QEMU to the guest. Memory copy is expensive for the packet processing.
- In the message path, when the packet arrives at the TAP device, the host kernel sends a notification message to QEMU. QEMU then requests KVM to issue interrupt using IOCTL mechanism, and then KVM sends the interrupt to the guest. This path incurred unnecessary performance overhead.

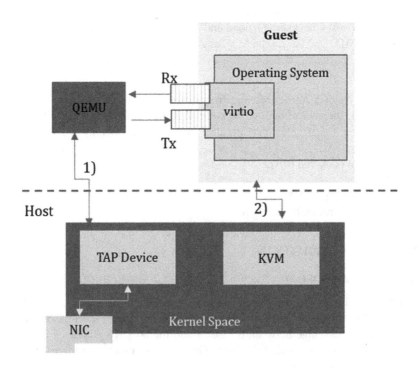

FIGURE 13.1   Before vhost-net.

13.1.2 VHOST-NET

Vhost-net kernel module is designed to optimize this; the main enhancement is to reduce the QEMU burden. Figure 13.2 describes the vhost-net system architecture.

- In the datapath, the packet data received from the TAP device is copied to the area which is owned by the virtual queue through the vhost-net module and the guest can receive the packet with one memory copy.
- In the message path, when the packet from the TAP device arrives at vhost-net, an interrupt is sent to the guest through the KVM module and the guest is notified. This reduces the QEMU processing (Figure 13.2).

The packet transmission process is similar, no need to repeat. The Linux kernel-based vhost-net has the following design assumptions on QEMU.

- QEMU shares the guest memory layout. Vhost-net is allowed to access the address translation information from guest physical address (GPA) to host physical address (HPA).
- QEMU shares the address of the virtual queues. Vhost-net can access them for packet sending and receiving. Because the virtual queue address is the address in the virtual space of the QEMU process, it should be translated into the virtual address in the process where vhost-net resides.

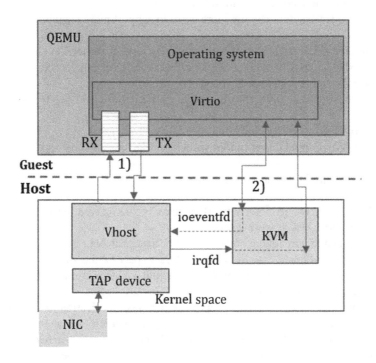

FIGURE 13.2   Vhost-net architecture.

- QEMU shares the event file descriptor (eventfd), which is configured in KVM for sending interrupts to the vhost-net device in the guest. In this way, vhost-net can notify the guest to receive the arrived packet in the receive queue.
- QEMU shares the eventfd which is configured in KVM, and the guest can write to vhost-net PCI configuration space. The event will be triggered when there is a guest to write to PCI configuration space on the vhost-net port. This method can notify vhost-net in the host that there is a packet to be sent.

13.1.3 VHOST-USER

The user space version of vhost-net, known as vhost-user, is supported by DPDK. The design goal is to increase the networking throughput for VM. Prior to DPDK, a NIC driver is running in the kernel space and the vhost-net is also a kernel module. The whole packet path moves from the NIC driver to software switch and then to guest; all can happen in Linux kernel efficiently.

DPDK introduced the PMD for high-speed NIC interface, and the idea is to move the whole software switch to the user space. Similar to vhost-net, vhost-user implements the virtual queue, message notifications. In addition, vhost-user can run the poll mode with the dedicated cores assigned. This aligns well with two major industry movements:

- More and more cores are built into CPU.
- Cloud infrastructure can use more cores to support more tenants.

As a result, the networking is a significant portion as packets need to be sent to tenants (guests). The dedicated cores are an effective way for efficient resource usage and deliver the stable system performance.

DPDK vhost-user supports the guest to run any kind of virtio-net driver.

- Linux virtio-net driver in kernel space.
- DPDK virtio-pmd driver in user space.

Vhost-user has the following basic functions:

- Management of virtio-net network devices, including creation and destruction of virtio-net network devices.
- Mapping and unmapping of the descriptor list, available ring, and used ring in the virtual queue into the virtual address space of the vhost process, as well as mapping and unmapping of the actual packet buffer into the virtual address space of the vhost process.
- When a packet is received, it sends the message notification to the guest; when a packet is sent, it receives a message notification from the guest.
- Support the packet exchange between virtio-net devices (via virtual queues, guest to guest) and between the virtio-net device and physical device (via NIC hardware queues).

- Support the jumbo frame with multiple merged buffers.
- Support the multi-queue to achieve performance scaling.

13.1.4 Vhost-User Acceleration (vDPA)

Vhost datapath acceleration (vDPA) is the latest design in implementing the vhost datapath on accelerators such as FPGA-based Smart NIC. This approach continues supporting the virtio interface, which is a proven standard interface. So the guests can stay unmodified, the accelerator can move packets using DMA, and this is beyond what a standard NIC can deliver. This architecture is a good candidate for NFV scenario, virtio-based interface address the key challenge, the standard interface can ensure the inter-operable approach across multiple vendors. This approach unifies the NIC interface from multiple vendors.

13.2 VHOST-USER DEEP DIVE

Vhost-user must support many guests simultaneously. It creates, manages, and destroys the vhost devices for the corresponding virtio-net devices running in the guest.

13.2.1 Message Mechanism

Vhost-user is based on a socket server to process the QEMU's messages. If there is a new socket connection, the guest creates a new virtio-net device and the vhost-user driver will create a vhost device in the back-end to pair with. When the socket connection is closed, vhost-user will destroy the corresponding back-end device instance. The detailed message interaction between the QEMU and vhost is shown in Figure 13.3.

VHOST_GET_FEATURES
Get a subset of virtio-net features and allow vhost as a back-end to make the decision.

VHOST_SET_FEATURES
Check the virtio-supported features and enable any features that are supported by both sides: vhost and virtio-net.

VHOST_SET_OWNER
Set the owner of the device with the current process.

VHOST_RESET_OWNER
The current process releases the ownership of the device.

VHOST_SET_MEM_TABLE
Set the memory space layout information for address translation in the data buffer during packet sending and reception.

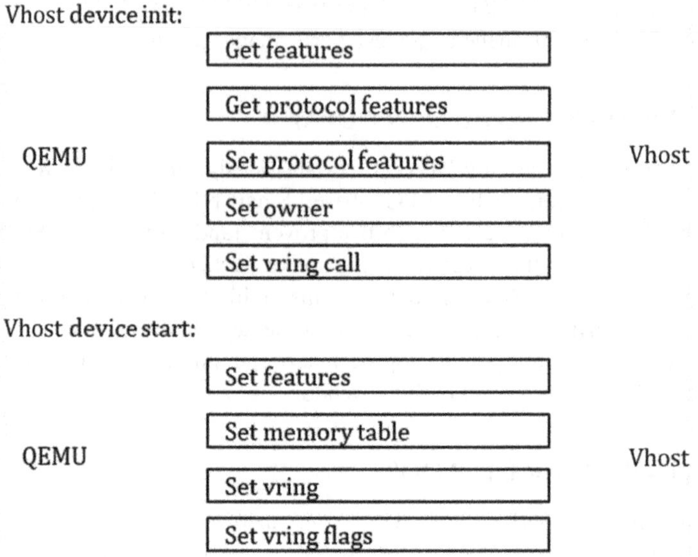

Vhost device init:

Vhost device start:

FIGURE 13.3   The messages between vhost and QEMU.

VHOST_SET_LOG_BASE/VHOST_SET_LOG_FD

The message can be used for guest online migration.

VHOST_SET_VRING_NUM

Vhost records every piece of information about every virtual queue (including receiving queue and sending queue).

VHOST_SET_VRING_ADDR

This message sends the virtual address of the virtqueue structure in the QEMU address space. Vhost translates this address into the vhost virtual address space. It uses VHOST_SET_VRING_NUM message to determine the sizes of the descriptor queue, avail queue, and used queue. (One page is generally assigned to each queue.)

VHOST_SET_BASE

The message delivers the initial index value, according to which vhost can find an available descriptor. Vhost also records the index value and sets it to the current location.

VHOST_GET_BASE

This message will return the current index value of vhost, i.e., the place where vhost expects to find a currently available descriptor.

VHOST_SET_VRING_KICK

This message delivers the eventfd. When a guest has a new packet to be sent, the file descriptor can be used to tell vhost to receive the new packet and send it to the destination. Vhost maps this eventfd from QEMU context to vhost process context by using an eventfd agent module.

VHOST_SET_VRING_CALL

This message also delivers an eventfd, which makes vhost receive the new packet and tell the guest to be ready to receive the new packet. Vhost maps this eventfd from QEMU context to vhost process context by using an eventfd agent module.

VHOST_USER_GET_VRING_BASE

This message sends the currently available index value of the virtual queue to QEMU.

13.2.2 ADDRESS TRANSLATION

QEMU supports a configurable option (mem-path) for routing directory/file system. QEMU allocates the required memory space in the file system. Therefore, it must ensure that there is enough memory space on the host and to reserve the memory earlier. (mem-prealloc).

In order to allow vhost to access virtual queues and packet buffers, certain pages must be mapped into the vhost process space. Those pages are used for the descriptor table, avail ring, and used ring of all the virtual queues.

After receiving the message VHOST_SET_MEM_TABLE from QEMU, vhost maps the QEMU physical memory into its own virtual memory space by using the memory table (file descriptor, address offset, block size, and other information) in the message.

Guest physical address (GPA): The guest physical address, such as the packet buffer address in a virtual queue, can be considered as an offset for returning to the starting address by using the above system function MMAP.

QEMU virtual address (QVA): QEMU delivers the location of the virtual queue in the QEMU virtual address space when it sends the message VHOST_SET_VRING_ADDR.

Vhost virtual address (VVA): To find the corresponding addresses of the virtual queue and the packet-contained buffer in the virtual address space of the vhost process, QVA and GPA must be translated into VVA.

In DPDK, the `rte_vhost_mem_region` data structure is used to store the region information and mapping relation of QEMU memory files. The overall region information is stored using the `rte_vhost_memory` data structure.

```
/**
 * Information relating to memory regions including offsets to
 * addresses in QEMUs memory file.
 */
struct rte_vhost_mem_region {
```

```
        /**< Base guest physical address of region. */
        uint64_t guest_phys_addr;
        /**< Base qemu virtual address of region. */
        uint64_t guest_user_addr;
        /**< Base vhost virtual address of region. */
        uint64_t host_user_addr;
         /**< Size of region. */
        uint64_t size;
         /**< Mmap address of this region. */
        void     *mmap_addr;
         /**< Mmap size of this region. */
        uint64_t mmap_size;
         /**< File descriptor of this region. */
        int fd;
};

/**
 * Memory structure includes region and mapping information.
 */
struct rte_vhost_memory {
        /**< Number of memory regions. */
        uint32_t nregions;
        /**< Memory region information. */
        struct rte_vhost_mem_region regions[];
};
```

With these two data structures, DPDK is able to calculate VVA corresponding to GPA or QVA through the address offset.

```
struct virtio_memory_regions *region;
vhost_va = guest_pa - region->guest_phys_addr +
               region->host_user_addr;
vhost_va = qemu_va - region->guest_user_addr +
               region->host_user_addr;
```

13.2.3 FEATURE NEGOTIATION

When initializing the device, the guest virtio-net front-end driver will query the features supported by the vhost back-end driver. After it receives the reply, it compares and computes the fields representing the vhost features and self-supported feature fields to determine the features requiring support from both and finally sends the available feature set to vhost.

The following is a set of features supported by DPDK vhost:

- VIRTIO_NET_F_HOST_TSO4: Host supports TSO V4.
- VIRTIO_NET_F_HOST_TSO6: Host supports TSO V6.
- VIRTIO_NET_F_CSUM: Host supports checksum.
- VIRTIO_NET_F_MRG_RXBUF: Host may merge packet receive buffers.
- VIRTIO_NET_F_MQ: Supports virtualization of multi-queues.

- VIRTIO_NET_F_CTRL_VQ: Supports control channel.
- VIRTIO_NET_F_CTRL_RX: Supports the receive-mode control channel.
- VHOST_USER_F_PROTOCOL_FEATURES: Supports vhost-user protocol feature negotiation.
- VHOST_F_LOG_ALL: Used for vhost live migration.
- VIRTIO_F_RING_F_EVENT_IDX: Supports event suppression.
- VIRTIO_F_IN_ORDER: Vhost will return descriptors in ring order.
- VIRTIO_F_IOMMU_PLATFORM: Supports virtual IOMMU.
- VIRTIO_F_RING_PACKED: Supports packed virtqueue.
- Feature bits are also introduced to allow vhost master and slave to negotiate the features. DPDK vhost supports the following vhost-user protocol features:
- VHOST_USER_PROTOCOL_F_MQ: Supports multiple queues.
- VHOST_USER_PROTOCOL_F_LOG_SHMFD: Supports VHOST_SET_LOG_BASE message.
- VHOST_USER_PROTOCOL_F_RARP: Supports VHOST_USER_SEND_RARP message.
- VHOST_USER_PROTOCOL_F_REPLY_ACK: Slave is able to reply success and failure.
- VHOST_USER_PROTOCOL_F_NET_MTU: Supports VHOST_USER_NET_SET_MTU message.
- VHOST_USER_PROTOCOL_F_SLAVE_REQ: Supports the slave channel.
- VHOST_USER_PROTOCOL_F_SLAVE_SEND_FD: Supports sending fds to master via slave channel.
- VHOST_USER_PROTOCOL_F_HOST_NOTIFIER: Supports setting memory-based host notifiers.
- VHOST_USER_PROTOCOL_F_PAGEFAULT: Supports postcopy migration.

13.2.4 Virtio-Net Device Management

The life cycle of a virtio-net device includes four stages: device creation, configuration, service kickoff, and device destruction.

1. *Device creation*: Vhost-user creates a device by establishing a socket connection.
   When you create a virtio-net device, you may need to:
   - Assign a new virtio-net device architecture and add it into the virtio-net device list.
   - Assign a processing core for serving the virtio-net device and add the virtio-net device into the list on the data plane.
   - On the vhost, assign a RX/TX queue for serving the virtio-net.
2. *Configuration*: Use the message VHOST _ SET _ VRING _ * to configure the size, basic index, and location of the virtual queue. With this information, vhost maps the virtual queue into its own virtual address space.
3. *Service kickoff*: Send the message VHOST _ USER _ SET _ VRING _ KICK to kick off the virtual queue service. Afterward, vhost can poll its receiving queues and put data in the receiving queue of the virtio-net

device. It can also poll the virtual sending queues to check for data packets ready to be sent. If there are any, they are copied to the sending queue.

4. *Destruction of virtio-net device*: Notify the service is stopped. After receiving the message, vhost will immediately stop the polling of the transmitted virtual queues and the polling of NIC receiving queues. At the same time, both the processing cores assigned to the virtio-net device and the RX/TX queues on the physical NIC will also be released.

13.2.5   Vhost Checksum Offload and TSO

To reduce the CPU overhead arising from the high-speed network system, most modern NICs support several feature off-loading techniques, as described in Chapter 9. Two important features are checksum offload and TSO (TCP segment offload).

- Checksum is widely used in the network protocol for verifying whether there is an error in the messaging process. If a NIC supports the checksum offload, the checksum calculation can be made in the NIC without consuming any CPU resources.
- The TSO (TCP segment offload) technique uses the NIC processing capability to divide a large TCP packet coming from the upper layer into several small TCP packets and finish tasks such as adding IP packet header, copying TCP header, and calculating the checksum of each small packet. Therefore, if a NIC does not support TSO, the TCP software protocol layer considers MSS (maximum segment size) when sending packets to the IP layer. A large packet will be divided into several packets and sent, leading to more CPU loads.

In DPDK vhost implementation, both the checksum offload and TSO can be accelerated by the NIC. The packet is sent from the guest to the host through the virtio using memory copy, and no physical network is involved. Therefore, the risk of packet transmission error can be ignored and it is not necessary to validate the packet checksum, nor segment the large-sized packets. Regarding the checksum offload and TSO support in vhost, it is okay to tell VM that these features have been supported during feature negotiation phase (based on NIC feature status). When the VM sends the packet via virtio-net, the packet is marked that the checksum in the packet header and TCP segmentation need to be completed in the vhost back-end. Finally, when vhost receives the packet, it can decide to leave the computation to NIC (when packet needs be sent out).

13.3   VHOST PMD

Vhost PMD driver is easy to use. The first thing is to register as the network device. Its data are structured as follows:

```
static struct rte_vdev_driver pmd_vhost_drv = {
        .probe = rte_pmd_vhost_probe,
        .remove = rte_pmd_vhost_remove,
};
```

The rte _ pmd _ vhost _ probe() calls eth _ dev _ vhost _ create() to register it.

rx _ pkt _ burst and tx _ pkt _ burst are the device functions for receiving and sending the packets.

Their registration in the vhost PMD device is as follows:

```
eth_dev->rx_pkt_burst = eth_vhost_rx;
eth_dev->tx_pkt_burst = eth_vhost_tx;
```

Upon completion of the virtual device registration, there is no difference between the vhost PMD virtual and any physical Ethernet device. The following code is a simple forwarding process from user perspective.

```
struct fwd_stream {
    portid_t   rx_port;    /* Packet receiving port */
    queueid_t  rx_queue;   /* Packet receiving queue */
    portid_t   tx_port;    /* Packet sending port */
    queueid_t  tx_queue;   /* Packet sending queue */
};
struct fwd_stream *fs;
/* Receive the packets from the receiving port */
nb_rx = rte _ eth _ rx _ burst(fs->rx_port, fs->rx_queue,
pkts_burst,
                        nb_pkt_per_burst);
if (unlikely(nb_rx == 0))
        return;
/* Send the packets via the sending port */
nb_tx = rte _ eth _ tx _ burst(fs->tx_port, fs->tx_queue,
pkts_burst, nb_rx);
/* If it fails, release the buffer */
if (unlikely(nb_tx < nb_rx)) {
        do {
                rte_pktmbuf_free(pkts_burst[nb_tx]);
        } while (++nb_tx < nb_rx);
}
```

Finally, rte _ eth _ rx _ burst and rte _ eth _ tx _ burst will, respectively, call rx _ pkt _ burst and tx _ pkt _ burst so that the vhost PMD virtual device will work on the packet RX/TX.

Multi-queues can be enabled in vhost-PMD with devargs "queues=<int>". The default number of queue pairs is 1. With multi-queues enabled, the application can use these queues with the standard queue-aware DPDK ethdev APIs.

13.4 CONCLUSION

Virtio and vhost are important software modules for KVM/QEMU-based virtualization, which are widely adopted in cloud computing, together with virtual switching software, which is responsible for the network virtualization/overlay realization.

Virtual switch will be explained in the later chapters. DPDK adds the new ingredients such as user space, poll mode, multi-queues and scales up the performance with multi-cores. Vhost-user PMD in DPDK gained the wide adoption in both cloud and NFV infrastructure today.

By adopting virtio-user to support the container guest, the same virtual switch infrastructure solution can be used for container-based network functions.

Section 3

DPDK Applications

This section focuses on how to use DPDK (Data Plane Development Kit) and other open-source projects; the focus is shifted to build the real-world applications.

- Chapter 14 will introduce DPDK, SDI (software defined infrastructure), and NFV (network function virtualization). The NFV evolution and VNF (virtual network function) optimization methods are explained here.
- Chapter 15 will introduce vSwitch and DPDK acceleration. The OVS (open vSwitch) data path architecture is discussed on how to use DPDK to get more performance.
- Chapter 16 will introduce DPDK to optimize the storage workload. This chapter will also introduce SPDK (Storage Performance Development Kit), an open source contributed from Intel.

14 DPDK for NFV

Xuekun Hu
Intel®

Waterman Cao
Huawei

Heqing Zhu
Intel®

CONTENTS

SDI means the software-defined infrastructure; a networking system can be implemented with software running on COTS. Big public cloud companies have large engineering talents, and those developers have built in-house networking systems for a few reasons:

- To simplify the network systems.
- To meet the custom needs of the network.
- To reduce the networking cost.
- To address the rising network virtualization and security needs (Figure 14.1).

The in-house development model is known as Cloud SDI. Over the years, many cloud companies have gone down this path to build network appliances such as virtual routers, firewalls, load balancers, VPNs on their own. VNF implies the virtual

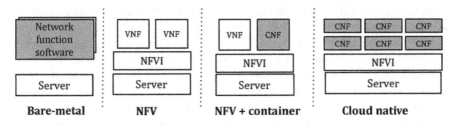

FIGURE 14.1 Cloud SDI and NFV model.

network function that runs within a VM, and CNF implies the network function that runs within a container instance. In these software-based network functions, DPDK is used as a high-performance I/O module. Cloud developers have a complete control over the system, and virtualization is not a must-have technology for in-house networking systems used in cloud. Figure 14.1 describes the multiple ways of building the network functions at cloud; the platform infrastructure can be bare-metal, virtualized, or containerized.

If we look at the cloud computing service to enterprise IT customers, the networking service is a premium software service running on cloud data center. Cloud providers offer multiple options to serve the customers with the online marketplace (such as AWS and Azure).

- A marketplace supports any third-party network functions (vRouter, vFirewall, etc.).
- Cloud builds in-house network functions (AWS-built virtual private network, Transit Gateway, etc.)

To run the third-party virtual network/security functions on cloud data center server, it requires the use of virtualization technology (or container technology as the underlying infrastructure) and it is subject to cloud provider's specific infrastructure and software framework. This can be an open opportunity for anyone who wants to deliver its software in the marketplace, and it just needs to run on the cloud infrastructure. So this model opens up the business opportunity for any software developers (and companies).

NFV (network function virtualization) concept is largely promoted by the telecom operators. The business motivation is similar to cloud model: applying software-based network functions on the general server platforms. When the concept emerged, the virtualization technology is the dominant choice of cloud infrastructure and the container technology has not been known as today (2019).

DPDK is chosen as a critical software ingredient to be part of NFV concept and system architecture, and it can be used for VNF (virtual network function), CNF (container network function), and NFVI (NFV infrastructure) solutions.

Container technology and cloud native concept have emerged since 2015; they became the preferred choice for cloud and IT service provision; and around the container, there a new wave of technology development on how network function is delivered (packaged) and deployed. Figure 14.1 shows the software evolution path for the network systems. The telecom operator expects to see decoupled HW/SW systems, this model will bring the flexibility to choose different HW vendors. From the software perspective, NFVI, VNF, and CNF can come from the different vendors as well. Because of the supplier complexity, NFV is a more complicated model than cloud SDI. As a result, the NFV adoption pace is slower than cloud SDI.

Cloud native architecture is a way to deliver the service at the micro-level: If a new service request comes, the elastic model suggests the new instance can be spawned as necessary, and there is no need to reserve the resource for peak capacity beforehand. Traditionally, the telecom system tends to reserve the maximum resource for peak performance scenario.

14.1 NETWORK FUNCTION VIRTUALIZATION

In October 2012, the leading global telecom carriers met at SDN and OpenFlow World Congress in Germany. The first NFV white paper was published, and the document presents how to use the general server platform to provide the telecom network service. After that, European Telecommunication Standards Institute (ETSI) established the Industry Specification Group (NFV ISG) as an international standard organization to regulate and accelerate the process. The ultimate goal is to virtualize IT resources with software, deploying important network functions in a virtual manner. See http://portal.etsi.org/NFV/NFV_White_Paper.pdf for the white paper details (Figure 14.2).

The traditional network appliances for telecom carriers are expensive and complex. As shown in Figure 14.2, there are a variety of network appliances and each comes with a specific system design, different chips, different software, different network interfaces, different management and control interfaces, different form factors, from different vendors. Large telecom vendor will have to invest a large number of talents on product development, and the proprietary hardware appliances are limited by the long product development cycle, from chip design to system development to network trial.

NFV is designed to reduce many network appliances and consolidate them on the standard high-volume servers, SDN switches, and storage systems. On a standard server, a network function is given in a software form. It can be deployed as needed, migrated, and deleted in the different locations; software instances can be launched (more or less) to adjust the dynamic demand. And this software model is independent of the underlying physical servers.

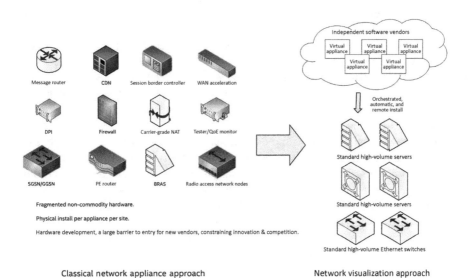

FIGURE 14.2 NFV transformation.

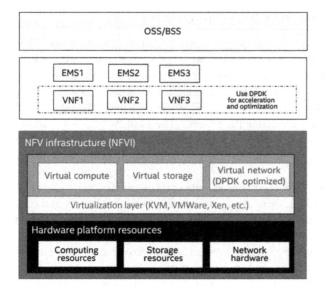
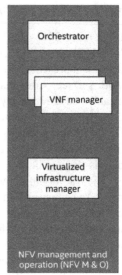

FIGURE 14.3   NFV framework.

NFV becomes technically feasible due to silicon readiness (the affordable CPU and high-speed NIC), cloud virtualization technology, and open sources such as Linux, KVM, DPDK, Open vSwitch, SR-IOV. NFV framework is published by ETSI in 2013 (Figure 14.3).

NFV has gone from the early concept to commercial adoption. In a recent China Mobile NFV bidding, the server platform and NFV software are the main purchase package. For telecom operator. It is a clear decouple strategy on selecting the hardware and software suppliers.

Take "Cisco NFV Infrastructure Architecture" as a real example. The infrastructure has the computing, storage, and network hardware. Cisco UCS are the server platform to support the computing and storage. The switch can connect all the server platforms, known as the physical network system. In addition to switch, there are the "virtualized network" and "virtualized network functions" are the software to implement a specific network function.

For more details about Cisco NFVI, please read further on its Web site: https://www.cisco.com/c/dam/en/us/solutions/collateral/service-provider/network-functions-virtualization-nfv-infrastructure/cisco-nfv-infrastructure-aag.pdf.

From the purchase perspective, a telecom operator can choose Cisco as NFVI vendor and then use Cisco or other suppliers to deliver the VNF solution. For example, the wireless operator need have the transitional path to update its 4G infrastructure. Vendor such as Cisco has come up a transitional software path to define/deliver/deploy the network infrastructure.

- Virtualization: This is the initial step to virtualize the existing network function.

- Control and user plane separation, this is one step further. The intent is to decouple the control and data plane from the system architecture perspective. This model enables the deployment on demand, and it opens up the open architecture and platform choices to implement the control plane or data plane functions.
- Take 5G non-stand-alone architecture as an example, for the initial 5G roll-out, one architecture is to place 5G processing on the existing wireless platform. And this is good for experiment trial at the early phase.
- Introduce network slicing. This is a flexible method. Using the resource on the basis of service demand, network slicing is the efficient way to edge service and increase the software service on demand.
- Generalize 5G and move to stand-alone architecture. While 5G service is transitioned into massive use, upgrade the 5G core network function and application integration with cloud-native implementation. This is still software upgrade with the open architecture and open platform.

For more details on the Cisco solution on moving from 4G EPC to 5G evolution path, please read the white paper on its Web site: https://www.cisco.com/c/dam/m/en_us/network-intelligence/service-provider/digital-transformation/pdfs/cisco-ultra-5g-packet-core-solution-wp-v1a.pdf.

VNF is a software-implemented network function. In the context of NFVI and VNF separation, the legacy network appliances are being significantly simplified using the the commodity server and switch platform. At the silicon level, the major components are CPU, switch chip and NIC, SSD, etc. DPDK is used on both NFVI and VNF.

There will be multiple VNFs running on a general-purpose server. Linux networking stack is not sufficient for I/O, compared to the compute workload, and VNF needs high-performance I/O; it is popular to run DPDK PMD in the guest (VNF). For a VNF running in a virtualization environment, the data will first enter the server platform and then be delivered to VNF. This packet delivery can be done by NIC or vSwitch. If the NIC is responsible for packet delivery to VNF, SR-IOV is a popular approach. If vSwitch is responsible for packet delivery, VNF can use the virtio-pmd as the virtualized interface and DPDK-accelerated vSwitch is a popular choice, such as OVS-DPDK. It will be described in the next chapter. Prior to DPDK acceleration, OVS (Open vSwitch) is a popular network virtualization software for cloud server.

14.2   VIRTUAL NETWORK FUNCTION

VNFs are software-defined network functions, for example firewalls, routers, or load balancers. VNF is a software module placed on a NFVI-capable server system. VNF placement need consider the following items:

- *System resources allocation*: The number of cores, memory, storage, and network interface should be evaluated. Multiple cores are often used as an easy way for performance scale; they pair with multiple queues for VNF.

Multiple queues can belong to a VF (NIC) or a virtualized network interface (virtio).

- *Virtualized interface*: Physical network interfaces are shared resources in a NFVI context. VNF can choose different interfaces: software interface (such as virtio), SR-IOV/VF interface PCIe device pass-through, or socket-based interface. When selecting the virtual network interface, it might be a trade-off when considering the performance and workload migration. In the guest, VNF can choose the Linux kernel networking with the standard socket interface. Or VNF can use DPDK PMD to accelerate the virtualized interface.

- *NIC driver*: Linux kernel module is the default driver for the installed NICs, and the I/O performance is not high with kernel driver. NIC can be partitioned with SR-IOV. Using VF with DPDK PMD can improve the network throughput to VNF or vSwitch.

- *Hardware acceleration*: Smart NIC, Intel® FPGA, and Intel® QAT are platform accelerators, and they can replace the software function to free up CPU. An accelerator can address the performance challenge, but it also introduces additional platform cost. VNF acceleration has the deployment challenge, and there is no abstracted accelerator interface that is known to be platform neutral. Without the software abstraction, the VNF acceleration causes the software to be dependent on the underlying hardware platform and this approach is not aligned with the hardware and software decouple principle.

- *QoS (quality of service) assurance*: Multiple VNFs share the platform resource. The consumption on the cache utilization and memory access frequency is workload specific. Intel® RDT technology can track and control the shared resource, including identifying the "noisy neighbor" to take actions to reduce performance interference.

From the telecom operators' perspective, VNF deployment also needs to consider the performance and system capacity requirement. If there is a need to increase the capacity,

- it can assign more system resources to the existing VNF instance,
- or launch more VNF instances to boost the system capacity.

From the performance characteristics, VNF is just software and it can be analyzed as compute intensive, memory intensive, or IO intensive.

- The main function of Router is routing and packet forwarding, and there will be intensive memory access and high I/O.
- vBRAS control plane function is mainly responsible for session management and authentication. It is compute intensive and has low packet rate. It is more about computation and low I/O.
- vBBU may include lots of digital processing, such as signal coding/decoding, and latency requirement is high.

It might be an easy decision for a cloud data center to assign a full rack of server to be used by vRouters and another rack of servers to be used for vFirewall. Then the service chain can be easily created between racks. The other model is to consider the smart VNF placement to make the best of platform resource (Deploying a service chain for vRouter and vFirewall in a same platform).

From VNF development perspective, the first step is to implement the network function with software and then move the software from non-virtualization to virtualization environment. The performance optimization can go with the methodology shown in Figure 14.4.

Deploy VNF in a non-virtualization environment, and then measure the performance metrics such as network throughput, latency, scalability. The performance tuning and optimization can be done with the closed-loop methodology: Do simple experiment and analyze possible bottleneck, find a place to optimize, and repeat the experiment to ensure the optimization is effective. System measurements need look at BIOS configuration and OS/kernel versions and configuration.

After establishing the performance metrics and desired resource allocation in a non-virtualization environment, deploy the VNF into a virtualization environment and repeat the method to hit a performance closer to the non-virtualization environment. Virtualization cost can be high because of the use of hypervisor and vSwitch; also, it is related to the guest operating system configuration.

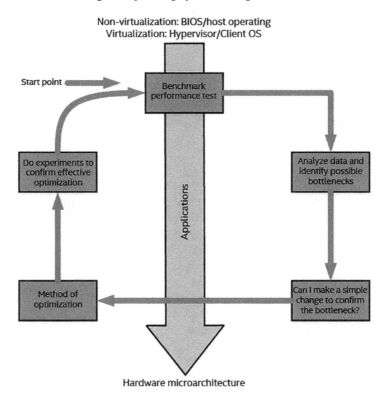

FIGURE 14.4   VNF performance analysis method.

Tables 14.1 and 14.2 list the performance tuning and optimization ideas on an Intel® server for a quick reference (Tables 14.1 and 14.2).

DPDK PMD is often selected by VNF to meet the high I/O needs, and this implies 100% core utilization. For the low I/O use case in the enterprise or cloud networking, Linux network stack might be sufficient enough.

14.3   OPEN-SOURCE RECIPE FOR NFV

When an enterprise start to use the cloud computing infrastructure, it can move all services to the public clouds such as AWS and Alibaba Cloud, or it can deploy the partial service on public cloud and partial service with on-premises cloud infrastructure at its own data center. Network and security functions can be purchased at the cloud marketplace online. Using the AWS marketplace as an example, F5 may offer the "BIG-IP" virtual edition with a performance range from 25 Mbps, 200 Mbps, 1 Gbps, 5 Gbps, 10 Gbps for application and service load balancing, so do many other companies that can deliver virtual network/security services in the marketplace, including AWS itself.

AWS virtual networking interfaces are publicly available and offer the choices of ENA (AWS's virtual NIC) and ixgbe interface. Both ENA and ixgbe PMD are available at DPDK Web site. Both ixgbe and ENA have the Linux kernel driver. ENA is a specific networking interface for AWS, and other cloud companies have their own.

From VNF perspective, `dpdk/examples/l3fwd` can be a simple example on how to get started with DPDK APIs. For years, DPDK has designed the abstract APIs. A VNF developer can depend on the DPDK APIs without any actual code change to configure the underlying NIC.

DPDK is not sufficient to build the network function, and the other recipes, such as open-source projects, can be helpful. There are many publicly available open-source networking or security projects. Here are just some famous examples, for the new developer to get started from somewhere, they can avoid doing something from scratch.

14.3.1   FD.io

FD.io is an open-source project under Linux foundation networking project. It includes a few sub-projects. VPP is one of the most significant projects, and it originated from Cisco. Overall, this open-source project has widely been used for commercial adoption for VNF implementation.

- *Cisco*: ASR 9000 Router, Carrier-Grade Service Engine.
- *Alibaba Cloud*: uCPE, load balancer.
- *Netgate*: TNSR, secure networking software.
- *Yahoo Japan*: cloud load balancer
- *ZTE*: IOT gateway, mobile edge computing platform, PaaS, etc.

FD.io/VPP is "vector packet processing". It is running in the user space, and it can be used for bare-metal, virtualized, or containerized environment. VPP is based on DPDK (and other I/O modules), the vector of packets can come from DPDK PMD, and VPP reads a bunch of packets with graph processing framework. A vector of

TABLE 14.1

Performance Tuning Perspective

Pain Point	Observation	Platform	OS	SW
Processor computing power	CPU utilization rate	Choose the high-frequency processor; choose high-core-count processor	Assign more cores, core-pinning	Multi-threaded; SIMD optimization
Memory	Memory bandwidth usage	Use high-bandwidth memory, interleave population at all channels	Assign more memory; enable hugepages	Use hugepages
Physical network I/O	Network throughput	More ports, high-speed interface; Intel® DDIO; NIC-based multi-queues	PF/VF, SR-IOV configuration	PMD
Network latency	Meet the need?	Local socket usage	Local resource usage (core, interrupt, etc.)	PMD, packet burst
VNF performance	Compare with bare-metal provision	Resource assignment	VM/NUMA aware	Tuning
NFVI	Number of VMs, virtual interface, etc.	Resource assignment	VT-d, VT-x	OVS-DPDK?

TABLE 14.2

VNF Infrastructure—Network Interface Options

Interface	Performance	Flexible Deployment	Maintenance
Virtio	Medium	High	Mature with community support
SR-IOV/VF	High, shared use by multiple VNFs	Low	Vendor support
NIC pass-through	Highest, exclusive use	Low	Vendor support
AF_XDP	High with AF_XDP PMD. Medium with socket(AF_XDP)	High, but need new Linux kernel	Need NIC driver ready

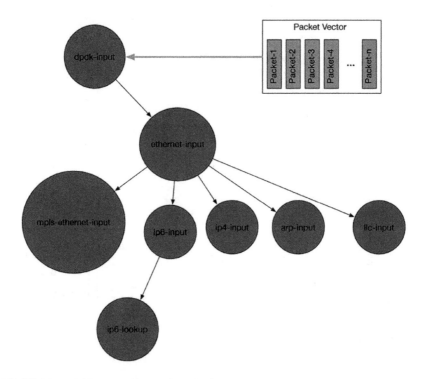

FIGURE 14.5   VPP concept in graph node view.

packets are processed together on a graph node. The normal software only processes one packet at a time. The vector packet processing is more efficient to make better use of CPU execution logic. The graph node is pluggable, and this supports the easy customization on the packet processing path (Figure 14.5).

VPP can be integrated with OpenStack with "Networking VPP" or "Open Daylight". FD.io can be integrated with Kubernetes for container provision with "Ligato" and "Contiv/VPP". This helps to implement the network function into the container-based architecture.

DPDK did not implement the TCP/IP protocol stack. VPP built the rich network protocol processing features such as MPLS, IPv4/IPv6, ARP, ECMP, DHCP, IPsec, ACL, network virtualization. Currently, the user space TLS and universal DPI development are happening as well. DPDK and VPP provide a flexible optimized packet processing framework and optimized implementation on top of SIMD, AVX/DDIO, and QAT. It demonstrated a 1 Tbps router on Intel® Purley Platform. VPP is also used for uCPE/vCPE/SD-WAN platform. It supports the network virtualization and service chaining. NSH-SFC is service chain project, it is a sister project under FD.io.

14.3.2 Snort

Snort is a popular open-source project from Cisco. Snort was an open-source project before Cisco acquired SourceFire in 2013. Snort is a widely used open source for intrusion detection and prevention system; Cisco NGIPS (Next-Generation IPS) is using Snort.

DAQ, "Data Acquisition Model", is the packet interface for Snort. DPDK can be used under Snort/DAQ module, and the other Intel® open source project "Hyperscan" can be integrated to accelerate Snort. Snort3 has the built-in support with Hyperscan acceleration.

Hyperscan is one of the fastest regular expression algorithms. Hyperscan is released at www.Hyperscan.io by Intel®. This open source can be used by other open-source security software such as Web application firewall (ModSecurity), deep packet inspection systems (nDPI), Suricata (another IDS/IPS) (Figure 14.6).

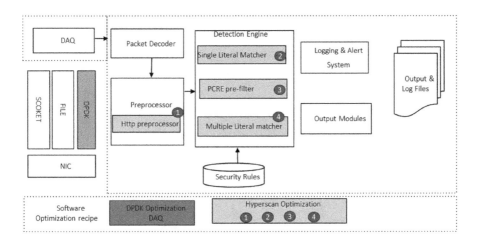

FIGURE 14.6 Snort optimization recipes.

15 Virtual Switch

Ciara Loftus, Xuekun Hu, and Heqing Zhu
Intel®

CONTENTS

The previous chapter described the concept of NFV and how network functions are being virtualized and run as applications in virtual machines (VMs). With this new architecture, the need for the virtual infrastructure layer that supports the communication arises:

- The datapath between these functions (VM to VM) in the same server is known as "east–west" traffic.
- The datapath between the virtualized function and the external functions is known as "north–south" traffic, which exists across the multiple servers.

The virtual switch plays an important role here; usually, it delivers the network overlay, routing, and security service. It is the foundational network infrastructure for cloud computing and the network function virtualization (Figure 15.1).

There are many open-source virtual switches including Open vSwitch (OVS), Snabb Switch, Lagopus, VPP, Tungsten Fabric. Commercial solution examples include VMware NSX-T, Juniper Contrail, Wind River Titanium Edition, 6WINDGate, and all of them have adopted the DPDK-alike technologies to deliver the network virtualization (overlay) in the server. This chapter introduces the concept of a virtual switch and use Open vSwitch as an example.

15.1 INTRODUCTION

The virtual switch works in much the same way as its counterpart, the physical switch. The virtual switch maintains virtual connections to multiple network interfaces, both physical and virtual. Those interfaces are monitored by the vSwitch using either a "polling" model or, less frequently, an "interrupt" model. In polling mode, the vSwitch indefinitely queries the interface for ingress packets. In interrupt mode,

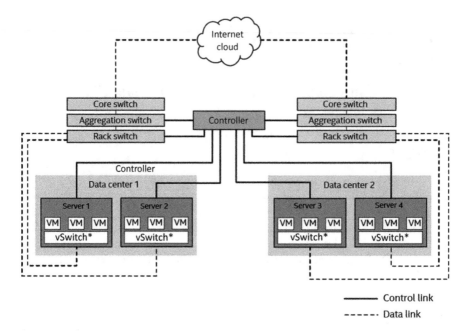

FIGURE 15.1   vSwitch role in the data center.

the vSwitch is notified via interrupt of packet ingression. Polling mode is more performant, but will utilize the full bandwidth available of the CPU on which it is being run. Typically, several cores will be used for polling the multiple interfaces.

Once the traffic reaches the vSwitch, it is processed and forwarded appropriately according to the flow rules which are programmed by the user or an external orchestrator. A basic example of a flow rule is given in Table 15.1.

The flow rule has two parts—a match and an action. The match part specifies some criteria that the ingress packet header must match. In the example below, the rule applies to all packets with IP destination in the range from 10.10.10.0 to 10.10.10.255. The range is specified by the * wildcard notifier. It could also be valid to specify a definite address, e.g., 10.10.10.10, or to specify additional match rules, e.g., "IP dest = 10.10.10.* and TCP dest = 443".

The action part of the flow rule specifies what to do with the ingress packets that match the specified criteria. The example shows a simple "output" action which would transmit the packet to the interface specified; in this case, the interface is associated with the port ID "2". vSwitches typically implement plenty other action

TABLE 15.1

vSwitch Flow Rule Example

Match	Action
IP Dest = 10.10.10.*	Output to Port 2

options, e.g., drop, apply VLAN tag, push/pop/modify header, and several actions can sometimes be chained together in the same flow rule.

The flow rules are stored in some form of a software flow table or hierarchy of flow tables. Hash tables are widely used for this purpose. Unlike a physical switch which is typically implemented by means of fixed function proprietary hardware with specialized high-speed memory, e.g., TCAM (ternary content-addressable memory), virtual switches can run on COTS hardware, making their deployment much more flexible, but resource limited. On platforms with limited memory, there can be challenges with storing and updating large sets of flow rules at a rapid pace. Searching these software flow tables will also take longer than it would for physical switches with TCAM, further reducing the speed at which packets can be switched in software compared to fixed-function hardware.

Some alternatives to virtual switching exist for north–south communication, such as SR-IOV and PCIe pass-through. Although very performant, these techniques are more simplistic and do not provide complex features such as connection tracking and network overlay (tunneling) that the virtual switch provides and, as a result, are not always suitable for real-world use cases. Most cloud data center infrastructure uses the vSwitch for workload provision.

15.2 OPEN vSWITCH

Open vSwitch (OVS) is an open-source virtual switch that is licensed under the Apache 2.0 license. The project is part of the Linux Foundation and at the time of writing has a 6-month release cadence, with releases typically coming in February and August of every year. It is compliant with the OpenFlow standard.

15.2.1 OVS Components

Several components are packaged with OVS and are key to its operation. Some of the core components are described in more detail in this section (Figure 15.2).

ovs-vswitchd: ovs-vswitchd is a user space daemon that contains the majority of OVS' switching functionality. Switching can be performed solely by the daemon in the user space, or it can be performed in conjunction with the OVS kernel module.

openvswitch.ko: The OVS kernel module is available in Linux kernels 3.3 and later and also comes packaged with Open vSwitch. It allows the user to perform packet classification and processing in kernel rather than in user space; however, some packets known as "exceptions" must be sent to user space if a match is not found in the kernel flow table. Section 15.2.2.1 elaborates on this architecture.

ovs-appctl: ovs-appctl is a tool used to control the ovs-vswitchd behavior and query its setting.

ovs-ofctl: ovs-ofctl is a tool which communicates with ovs-vswitchd via the OpenFlow protocol and delivers information regarding flow configuration. It can also be used to report the current state of OVS.

ovsdb-server: OVSDB stands for the Open vSwitch Database Management Protocol. The ovsdb-server delivers configuration information to the ovs-vswitchd, and the ovs-vswitchd updates the ovsdb with statistics and other runtime information.

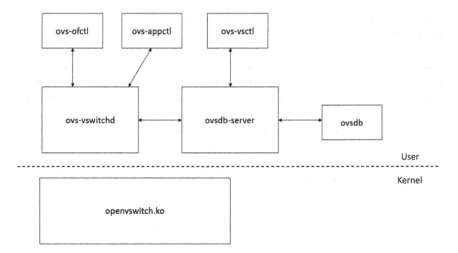

FIGURE 15.2 OVS components and relationships.

ovs-vsctl: ovs-vsctl is a tool that connects to the ovsdb-server and can be used to manipulate the configuration in the database.

Each of the OVS utilities has their own man(documentation) pages with extensive usage information.

15.2.2 OVS DATAPATH

As mentioned in the previous section, the OVS datapath can operate with or without the use of a companion kernel module (openvswitch.ko), at the user's discretion. This section discusses the architecture of each approach and how DPDK can be used to accelerate the user space approach.

15.2.2.1 Kernel (Native) Datapath

What is often referred to as the "native" architecture is that which includes the use of the Open vSwitch kernel module (openvswitch.ko). In this architecture, the majority of the packet processing and forwarding happens in the context of the kernel module (Figure 15.3).

Flow rules which have been configured in the user space through the tools and components mentioned in the previous sections are represented in the kernel in a simple flow table. Packets which ingress on a NIC are received by the kernel module; their headers are inspected and compared against the entries in the flow table and forwarded accordingly. In the case where a packet does not find a matching rule in the kernel flow table, this packet is known as an exception packet and it generates a message called an "upcall" which is sent to the user space daemon via the Netlink protocol. The relevant rule for the exception packet is found and propagated back to the kernel module such that the subsequent packets of the same format do not need to generate further upcalls and can be handled in kernel. The context switch from kernel to user space is costly in terms of CPU cycles.

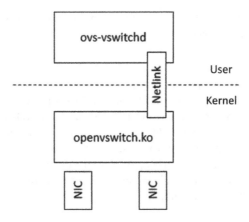

FIGURE 15.3   Internal module of OVS datapath.

15.2.2.2   User space Datapath and DPDK/PMD Acceleration

The user space datapath moves the forwarding plane into user space, removing the dependency on the kernel module and allowing all packet processing to be performed in user space. This architecture makes it possible to leverage DPDK's user space libraries and drivers for accelerated packet processing and IO, as indicated in Figure 15.4. With 4 optimization opportunities are introduced here.

Packets which ingress on DPDK PMDs are received by the user space OVS process, and their headers will be inspected and compared against a hierarchy of flow tables described in an upcoming section, with the exception packets eventually

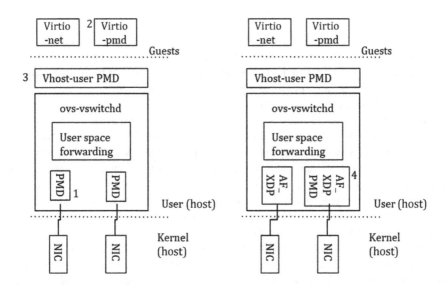

FIGURE 15.4   DPDK acceleration opportunities for OVS.

following a similar fate as those in the native architecture but without the need to cross the "user space"–"kernel space" boundary.

15.2.2.2.1 IO

Physical IO: NIC PMD (Opportunity #1)

In DPDK-accelerated OVS, physical IO, i.e., IO via physical NICs, is provided by means of DPDK Poll Mode Drivers (PMDs). These PMDs are added to OVS using the ovs-vsctl tool and the "add-port" command. The name, interface "type", and additional device-specific information are supplied as part of the command. For physical devices, their "type" is "dpdk" and the device-specific information we provide is the PCI address of the given device. Prior to adding ports, a bridge must be added using the "add-br" command.

```
ovs-vsctl add-br br0
ovs-vsctl add-port br0 dpdk0 -- set Interface dpdk0 type=dpdk
options:dpdk-devargs=0000:01:00.0
```

Physical IO: AF_XDP (Opportunity #4)

The other approach is to use AF_XDP netdev, which depends on Linux kernel driver with XDP/eBP/AF_XDP path or AF_XDP PMD (part of DPDK). AF_XDP can be added to OVS using the ovs-vsctl tool and the "add-port" command. The name, interface "type", and additional device-specific information are supplied as part of the command.

```
ovs-vsctl add-br br0 -- set Bridge br0 datapath_type=netdev
ovs-vsctl add-port br0 -- set Interface br0
datapath_type=af_xdp\
  options:n_rxq=1 options:xdpmode=drv \
  other_config:pmd-rxq-affinity="0:4"
```

This approach removes the dependency on using any specific DPDK PMD, and the underlying NIC driver need be ready with XDP/AF_XDP support. This is less performant than the DPDK netdev, but more vendor neutral. Go to http://docs.open vswitch.org/en/latest/intro/install/afxdp/ to find more details.

Virtual IO: (Opportunity #2, #3)

Virtual IO refers to IO via virtual networking devices. DPDK offers several virtual PMDs or "vdevs", some of which can be used with OVS. However, the most common type of virtual IO used with DPDK-accelerated OVS is user space vhost. DPDK-accelerated OVS has integrated the DPDK vhost library and offers ports of type "dpdkvhostuser" and "dpdkvhostuserclient" which act as back-ends to a virtio-net device which resides in a VM or container. The vhost library can manipulate the virtio-net rings directly, enabling the transmission and reception of packets to and from virtio-net devices (Figure 15.5).

The difference between dpdkvhostuser and dpdkvhostuserclient ports lies in with whom the responsibility lies for the creation of the socket which is used for control path communication between the virtio-net device and the back-end application. In server mode, the application, in our case DPDK-accelerated

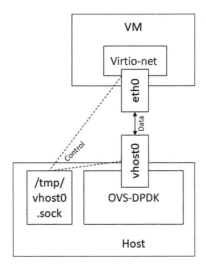

FIGURE 15.5   OVS-DPDK and vhost/virtio.

OVS, creates the socket, whereas in client mode, the server, e.g., the VM, creates the socket. Client mode is recommended as it means that if the application needs to be restarted the socket is not destroyed and once the app is restarted it can reconnect to the socket.

To add a vhost-user client mode device to OVS, the add-port command is used, the "type" is set to "dpdkvhostuserclient", and an additional parameter is supplied, which is the location of the socket.

```
ovs-vsctl add-port br0 vhost0 -- set Interface vhost0
type=dpdkvhostuserclient -- set Interface vhost0
options:vhost-server-path=/tmp/vhost0.sock
```

15.2.2.2.2   Classification

Packet classification is the process by which the vSwitch determines the correct course of action to take for a given ingress packet. At the time of writing, the default user space OVS datapath has a three-level hierarchy of flow tables which are used for the classification process:

1. EMC (exact match cache).
2. DPCLS (datapath classifier).
3. Ofproto.

The EMC contains a store of the most frequently seen flows in OVS. It has a capacity for 8192 entries, a size which was chosen for CPU cache size reasons. By default, packets which reach the switch will be first compared against entries in the EMC, using the RSS hash of the packet as an index into the table. A match will only be found if the packet header is an exact match with one of the header representations in the cache (Figure 15.6).

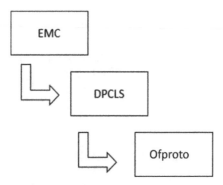

FIGURE 15.6   OVS-DPDK classification hierarchy.

If the packet fails to match in the EMC, it is compared against wildcarded flows in the DPCLS. The DPCLS is a tuple space search classifier (TSS) and has a capacity for 65,536 entries. The DPCLS itself is arranged in a series of sub-tables, each associated with a specific packet mask. If a match is found in the DPCLS, the EMC may be updated with the particular flow such that subsequent packets will not need to query the DPCLS.

If a match is not found in the DPCLS, the ofproto layer is queried. Again, if a match is found here, then the previous level in the hierarchy is updated with a new flow rule—in this case, a new rule is inserted into the DPCLS. Should this lookup fail, either the packet is dropped or sent to a remote SDN controller.

The classification performance is critical to the overall performance of any vSwitch, and OVS is no different. As such, different types of optimizations are continuously being researched and developed in order to improve the performance of each level in the hierarchy. Some examples of recent optimizations include the following:

- Probabilistic EMC insertion
    This limits the amount of new entries in the EMC to 1 in every potential 100 insertions. This theoretically should reduce thrashing of the EMC when there are a high number of parallel flows.
- Signature Match Cache (SMC)
    This is an additional level of cache added after the EMC that is more memory efficient than the EMC as it stores flow "signatures" rather than the full flow.

For basic setups, the defaults should suffice; however, for more complicated setups with particular traffic profiles and rates, it is recommended to trial the different options available (e.g., enable/disable EMC/SMC and change EMC insertion probability) in order to find the most optimal setup for the given use case.

Flows are added to OVS using the ovs-ofctl "add-flow" command, specifying the match and action criteria discussed in a previous section. Here, the match criterion is that the packet has ingressed on the port with ID 1 and the action is to output to the port with ID 2.

```
ovs-ofctl add-flow br0 in_port=1,action=output:2
```

15.2.2.2.3   Benchmarking and Testing
Basic setups for testing OVS with DPDK use "P2P", "PVP", and "VV" topologies, where "P" indicates a physical interface and "V" indicates a virtual interface. A port is added to the vSwitch for each interface in the setup, and a flow rule is added for each switch that must be performed to facilitate the desired traffic flow.

P2P (Physical to Physical)
The P2P topology simply forwards packets between a pair of physical ports. This is not a meaningful topology in terms of real-world use cases, but is often used to verify the basic operation of the switch before moving onto more complex use cases with virtual ports (Figure 15.7).

```
ovs-vsctl add-br br0
ovs-vsctl add-port br0 dpdk0 -- set Interface dpdk0 type=dpdk
options:dpdk-devargs=0000:01:00.0
ovs-vsctl add-port br0 dpdk1 -- set Interface dpdk1 type=dpdk
options:dpdk-devargs=0000:01:00.1
ovs-ofctl add-flow br0 in_port=1,action=output:2
ovs-ofctl add-flow br0 in_port=2,action=output:1
```

VP (Physical -> Virtual -> Physical)
The PVP topology forwards the traffic between pairs of physical and virtual ports, with an arbitrary workload in the VM that forwards the traffic between the two virtual interfaces. For example, the forwarding of traffic in the VM could be configured between a pair of virtio-net devices by using the Linux networking stack and configuring IP rules. Another option is to run a DPDK workload in the VM and bind the virtio-net devices to the DPDK virtio-pmd (Figure 15.8).

```
ovs-vsctl add-br br0

ovs-vsctl add-port br0 dpdk0 -- set Interface dpdk0 type=dpdk
options:dpdk-devargs=0000:01:00.0
ovs-vsctl add-port br0 dpdk1 -- set Interface dpdk1 type=dpdk
options:dpdk-devargs=0000:01:00.1
```

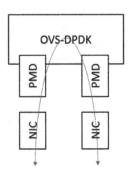

FIGURE 15.7   OVS-DPDK P2P setup.

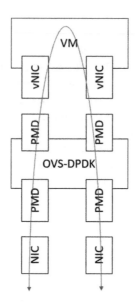

FIGURE 15.8   OVS-DPDK PVP setup.

```
ovs-vsctl add-port br0 vhost0 -- set Interface vhost0
type=dpdkvhostuserclient -- set Interface vhost0
options:vhost-server-path=/tmp/vhost0.sock
ovs-vsctl add-port br0 vhost1 -- set Interface vhost1
type=dpdkvhostuserclient -- set Interface vhost1
options:vhost-server-path=/tmp/vhost1.sock

ovs-ofctl add-flow br0 in_port=1,action=output:3
ovs-ofctl add-flow br0 in_port=4,action=output:2
ovs-ofctl add-flow br0 in_port=3,action=output:1
ovs-ofctl add-flow br0 in_port=2,action=output:4
```

VV (Virtual to Virtual)

The VV topology forwards the traffic between pairs of virtual ports in separate VMs. This type of traffic flow is often referred to as "east–west". The traffic is sourced in VM0, and this could be generated by means of a simple "ping" or through a more comprehensive tool like "iperf" (Figure 15.9).

```
ovs-vsctl add-br br0

ovs-vsctl add-port br0 vhost0 -- set Interface vhost0
type=dpdkvhostuserclient -- set Interface vhost0
options:vhost-server-path=/tmp/vhost0.sock
ovs-vsctl add-port br0 vhost1 -- set Interface vhost1
type=dpdkvhostuserclient -- set Interface vhost1
options:vhost-server-path=/tmp/vhost1.sock
```

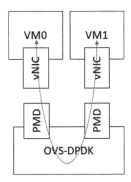

FIGURE 15.9   OVS-DPDK VPV setup.

```
ovs-ofctl add-flow br0 in_port=1,action=output:2
ovs-ofctl add-flow br0 in_port=2,action=output:1
```

The east–west traffic is becoming increasingly important, particularly for use cases such as service function chaining (SFC) where multiple network functions exist on the same platform and the traffic needs to be passed from one to the next in an efficient manner. NSH is a new header that according to the RFC "provides a mechanism for metadata exchange along the instantiated service paths". OVS provides support for the NSH header to facilitate SFC-type topologies; however, the configuration of such tests is out of the scope of this chapter.

15.2.2.2.4   Performance Tuning
Multi-queue

One way to improve the performance and scalability of the PVP test case, or indeed any test case using virtual ports is to increase the number of queues the virtual device has and thus increase its maximum bandwidth. Newer versions of OVS will automatically detect the number of queues on the front-end and reconfigure themselves accordingly.

For the PVP setup, it is recommended to configure the physical ports with the same amount of queues as the virtual ports and to configure the number of polling cores to that number as well. For example, if the virtual ports have two queues, configure the physical port queues and coremask like so:

```
ovs-vsctl set Interface dpdk0 options:n_rxq=2
ovs-vsctl set Interface dpdk1 options:n_rxq=2
ovs-vsctl set Open_vSwitch.other_config:pmd-cpu-mask=3
```

On the VM, if using the virtio-net driver, use ethtool to configure the queues:

```
ethtool -L eth0 combined 2
ethtool -L eth1 combined 2
```

If using a DPDK driver, if the option is available specify the number of RXQs and TXQs on the application command line like so:

```
./dpdk-app -c 0x3-n 4 … --rxq=2--txq=2
```

Finally, send the traffic in the physical ports that are spread across the queues via RSS. This can be achieved by configuring the input traffic with a range of different IP addresses or TCP/UDP port numbers. Here are the general tips.

- Ensure PMD threads (set by `pmd-cpu-mask`) are affinitized to the same NUMA node as that of the physical port(s).
- Ensure VM threads are affinitized to the same NUMA node as the PMD threads.
- Isolate cores which will run the PMD and VM threads in the kernel boot parameters using the "`isolcpus`" list.
- Enable compiler optimizations, e.g., "`-O3-march=native`".
- Assign multiple PMD threads especially if multiple ports are active.
- Enable hyperthreading or SMT.
- Configure EMC insertion probability "`emc-insert-inv-prob`" depending on the traffic profile.
- Configure output packet batching interval "`tx-flush-interval`" depending on the traffic load.

15.2.2.2.5 Performance Data

This section presents the performance data comparing OVS-DPDK with "native" OVS both v2.4.9 for the P2P and PVP test setups described above, running on a

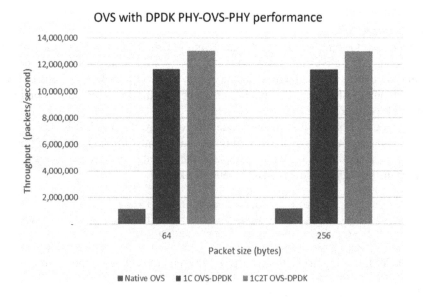

FIGURE 15.10 OVS-DPDK P2P benchmark.

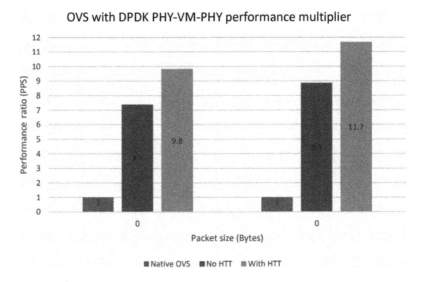

FIGURE 15.11   OVS-DPDK PVP benchmark.

platform with a 2.10 GHz Intel® Xeon® processor E5–2695 v4. Full test specification can be found on the www.01.org Web site in the ONP 2.1 Performance Test Report (Figures 15.10 and 15.11):

https://download.01.org/packet-processing/ONPS2.1/Intel_ONP_ Release_2.1_Performance_Test_Report_Rev1.0.pdf

15.3   CONCLUSION

In this chapter, the concept of a virtual switch and its architecture were introduced. It was shown how DPDK can be used to accelerate the performance of OVS and how that can be demonstrated using some basic benchmarking setups and practices. To achieve the best performance with DPDK-accelerated vSwitch, it is recommended to apply virtio-pmd in the guest, which is a common practice when building virtual network functions for higher I/O performance.

In addition to virtio-pmd for KVM-based virtualization, DPDK provides the other abstract software interfaces, such as vmxnet3-pmd for VMware NSX-T-based virtualized infrastructure. For Azure cloud and AWS cloud, virtual network interfaces are different and the PMD is also available in DPDK.

In recent year, a strong trend is to offload vSwitch to Smart NIC, which release cores for more tenant use. This model makes sense if the all Xeon cores are well utilizated. But if the system has lots of Xeon cores, many cores are not fully utilized by tenant workload. In this scenario, going with Smart NIC, implies the high platform cost.

16 Storage Acceleration

Ziye Yang and Heqing Zhu
Intel®

CONTENTS

DPDK can be applied to the storage system to improve the network efficiency. A network always plays a big role in the overall storage system performance. SPDK (Storage Performance Development Kit, http://spdk.io) is an open-source community-based effort, which focuses on the storage workload optimization. DPDK is an important ingredient in SPDK. This chapter shares a few pain points and solutions, and how DPDK is used to optimize storage workload.

16.1 THE STORAGE SYSTEM

Usually, a storage system has an internal network and an external network. The internal network is used for communication between nodes within a storage cluster. The external network is the external interface to represent the storage service. Ethernet is an integral part of today's storage system (Figure 16.1).

The network storage systems are connected with Ethernet network (NIC). Figure 16.1 describes an example. The storage system consists of multiple servers to provide different storage services. The common network protocols are all based on Ethernet. DPDK PMD can optimize the NIC performance, and it is faster than Linux kernel driver.

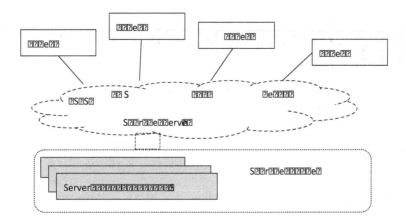

FIGURE 16.1   Network (Ethernet) storage system.

The storage services are listed below:

- iSCSI protocol: Provides the block storage service on SCSI (Small Computer System Interface).
- NAS protocol: Provides file storage service on Ethernet.
- Restful protocol: Provides object storage service on HTTP.
- FCoE (Fiber Channel over Ethernet): Provides storage service on the traditional SAN (storage area network).

iSCSI (Internet SCSI) is the network storage standard specification developed by IETF. It is used to connect storage devices through an Internet Protocol. By transferring SCSI commands and data over an IP network, iSCSI completes the data transfer between network nodes and enhances remote data management. The excellent data transmission capability of the iSCSI protocol is a key factor driving the SAN market. IP network is everywhere, and iSCSI can enable the data transfer on LANs, WANs, and even Internet, which removes the regional restrictions on storage service. iSCSI depends on TCP/IP network protocol, and SCSI command and data are encapsulated by iSCSI and TCP/IP packets, as shown in Figure 16.2.

Network storage system performance relies on the network performance and back-end storage system performance. System optimization need look at both places such as:

- Improve the front-end network performance.
- Improve the back-end data processing efficiency of the storage device.

If there is a write request from a user (see Figure 16.3), the request is processed as follows:

1. The NIC receives the data.
2. The data are copied from the kernel mode to user mode application.

FIGURE 16.2   iSCSI packet format.

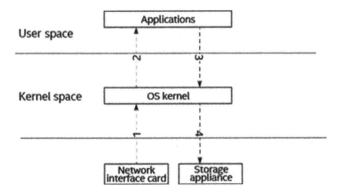

FIGURE 16.3   Write service for a storage system.

3. The application asks the kernel to write the data to the storage device. The data are copied from the user mode to kernel mode.
4. The driver in kernel completes the data write process to the storage device (Figure 16.3).

There are two data copies in this "data write" process of this network storage system. The first copy is caused by network-based request, and the second is caused by the back-end storage device. There are high performance storage systems that will optimize them with zero-copy approach. Linux supports the device with scatter–gather DMA engine, which may reduce the data copy from the kernel space to user space. It reduces the data movement cost between NIC and the application's memory space; the storage driver still needs to write the data to the device. Not all data copies can be avoided using Linux kernel storage system. How to improve the performance of the front-end and back-end is a challenge for network storage systems.

16.2   OPTIMIZATION OPPORTUNITIES

Figure 16.4 describes how to utilize DPDK to optimize the network storage system.

DPDK can optimize the NIC software driver. For instance, Intel® NIC supports the Flow Director feature, which can direct the TCP flow into the NIC queue and DPDK can create the RX/TX queue with TCP flow granularity.

The storage application accesses the network stack via socket interface. Linux kernel-based stack is feature rich, but less performant. An efficient user mode network protocol stack is desired to accelerate the network stack. In the ideal condition,

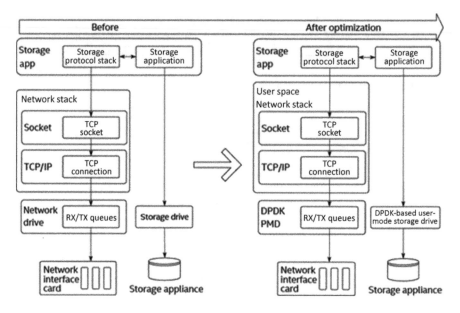

FIGURE 16.4 Optimization view.

all user space applications could be seamlessly migrated to use a high-performing network stack, without any code modification.

Prior to DPDK, the storage device driver is mostly kernel based. PMD concept can be applied here to implement the user space, poll mode storage device driver. This can boost data read/write throughput to the storage device.

16.3 SPDK (OPEN SOURCE)

Intel® leads an open-source project that focuses on IA platform-based acceleration library for storage workload, known as SPDK. It is a collection of software libraries to build high-performance storage systems. SPDK adopts lots of DPDK optimization ideas, and it includes the PMDs for NVMe SSDs. SPDK also has a set of environmental abstraction libraries (located in the lib/env directory) to manage CPU, memory, PCIe, and other device resources used by SPDK storage applications. DPDK is the default environment library for SPDK. Every new SPDK release depends on the latest DPDK stable release version.

The detailed architecture of SPDK is illustrated in Figure 16.5, including physical and virtualized PMD, application framework, storage devices, and storage protocols. SPDK can be integrated with many leading open-source projects such as Ceph, QEMU, RocksDB (Figure 16.5).

Storage protocols contain the accelerated applications implemented upon SPDK framework to support various protocols, e.g., iSCSI target service acceleration, vhost SCSI/blk in VMs, and NVMe-oF target over fabrics.

SPDK ARCHITECTURE

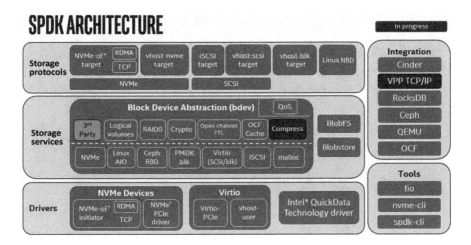

FIGURE 16.5   SPDK architecture.

16.3.1   APPLICATION FRAMEWORK

High-performing applications take advantage of multiple threads to support high concurrent connections. SPDK provides an event framework to address the data sharing among threads (Figure 16.6).

It adds a few concepts: **reactors, events,** and **pollers.** It spawns one thread per core (i.e., reactor), and each thread has lockless queues. Messages (events) can then be passed between the threads, and it is faster than the traditional locking mechanism.

- *Events*: Message passing is implemented in the form of events. The event framework runs an event loop thread per CPU core. These threads are called reactors which handle the incoming events from a queue. Each event is associated with a function pointer, with a destination core. Events are created using `spdk_event_allocate()` and executed by `spdk_event_call()`. The event-driven model requires the use of explicit asynchronous operation to achieve concurrency. Asynchronous I/O may be issued with a non-blocking function call, and completion is typically signaled with a callback function.
- *Reactors*: Each reactor has a lock-free queue for incoming events to the core, and any thread (core) can insert events into the queue of any other core. The reactor is a loop to handle the incoming events in FIFO order. Event function should never block and should execute very quickly.
- *Pollers*: It is a function registered with the `spdk_poller_register()`, which can be bound and executed on a specific core. Unlike events, pollers are executed repeatedly until it is unregistered. Pollers are intended to check hardware and are a replacement for interrupt-based mechanism. Normally, pollers are executed on main event loop and may be executed on a timer.

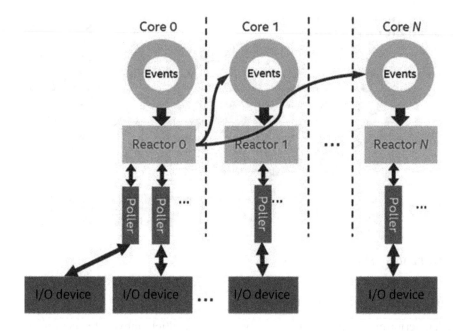

FIGURE 16.6   SPDK event framework.

16.3.2 Storage PMD

SPDK moves the storage device driver to the user space, using the poll mode, a similar DPDK concept for NIC device. Compared to Linux kernel-based driver, this removes the substantial overhead of system calls. As the storage application lives in the user space, this also eliminates the data copy between the kernel and user space.

- *Asynchronous I/O mode*: SPDK uses the asynchronous mode for I/O access request. SPDK implements the asynchronous read/write interface to allow the application to request the I/O, and a further function check is followed for I/O completion. When the I/O queue depth is larger than 1, such a methodology is more efficient than the synchronous mode. (CPU waits until I/O access completion.) Generally, it reduces the average I/O latency and increases the IOPS.
- *Lockfree design*: Traditionally, when multiple threads (cores) want to access the same NVMe device, the kernel driver needs to access the shared resource to avoid the data racing. SPDK adopts a lockless design, and each core (thread) accesses its own resources such as memory, I/O submission queue, I/O completion queue. Thus, I/O performance can be greatly improved. It is recommended that storage applications need to be changed to take advantage of this SPDK I/O framework. The change might not be trivial, and it has been proved with a good return with a significant performance benefit.

16.3.3 STORAGE SERVICE

Storage service is an abstraction layer with the block I/O interface to application. SPDK can export blockdevs (i.e., bdevs) constructed by NVMe drivers, Linux asynchronous I/O (libaio), Ceph RADOS block device API, etc. Storage applications can avoid directly working with low-level drivers. Also, blobstore/blobfs is implemented as user space file I/O interface, which can be integrated with RocksDB.

- *Blobstore* is a persistent, power-fail-safe block allocator, and it is designed to support a high-level storage service such as key/value stores (e.g., RocksDB) or local databases (MySQL), dedicated appliances (SAN and NAS), distributed storage systems (e.g., Cassandra). It is not intended to be a general file system, and it is intentionally not POSIX compliant. To avoid confusion, no reference to files or objects will be made at all, instead of using the term "blob". Blobs are typically large, measured in at least hundreds of kilobytes, and are always a multiple of the underlying block size. The blobstore is designed for asynchronous, uncached, parallel reads and writes on a block device.
- *BDEV* has the NVMe device abstraction and other libraries to export the block service interface to applications: API for implementing bdev drivers which enumerate and claim SPDK block devices and performance operations (read, write, unmap, etc.) on those devices; bdev drivers for NVMe, Linux AIO and Ceph RBD, blobdev (i.e., blobstore as bdev), Gpt bdev (i.e., construct bdevs upon existing bdevs via detecting GPT partitions); and configuration interface via SPDK configuration files or JSON RPC (Figure 16.7).

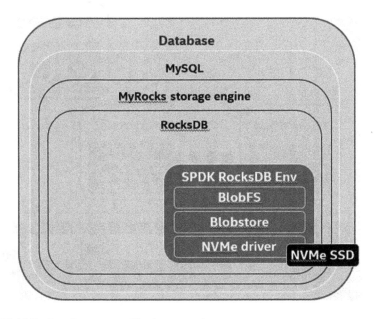

FIGURE 16.7   Local storage application example.

16.3.4 STORAGE PROTOCOLS

SPDK is suitable for the following three typical scenarios:

Back-end storage applications that provide block device interfaces such as iSCSI target and NVMe-oF target. These applications need support the iSCSI protocol or NVMe-oF protocols. These applications can be optimized according to SPDK framework, and the client application requires no change.

VM acceleration: If QEMU/KVM is used as a hypervisor, the vhost protocol implements an efficient vhost user space target based on a shared memory channel (for example, vhost SCSI/blk/NVMe target) to speed up the I/O while using the virtio SCSI/blk and kernel-native NVMe protocols in the virtual machine. It reduces the number of events such as VM interrupts (interrupt, VM_EXIT) and shortens the I/O stack in the host.

Database storage engine acceleration, such as RocksDB. SPDK's blobfs/blobstore can now be integrated to speed up the RocksDB engine to access NVMe SSD.

SPDK is not a universal solution for accelerating any storage applications yet. The user space stack and file system are the key development directions at the current development community. SDPK library can be selectively used to accelerate the specific storage applications.

16.3.5 PMD DETAILS

SPDK provides PMD for NVMe-based SSD device. NVMe is a protocol on top of PCIe interface (Figure 16.8).

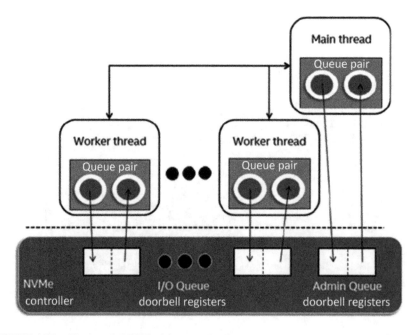

FIGURE 16.8 User mode NVMe driver mechanism.

1. NVMe PMD

SPDK adopts the PMD concept from DPDK and uses the similar CPU/ thread management for storage device's I/O access. It can avoid the interrupt and eliminate the context switching overhead. It can bind the I/O queue to the core (thread) for lockless access. SPDK uses the DPDK memory management techniques such as hugepages, memory buffer, pool to reduce the latency and increase core/cache utilization.

NVMe SSD is faster than the SSD using SATA and SAS protocols, and it has more IOPS and less latency. PMD saves the CPU consumption compared to the traditional kernel driver-based model. The experiment is configured as follows:

- The machine under experiment is loaded with 2x Xeon E5–2695v4.
- 64 GB DDR4 memory, 8x 8 GB DDR4 2133 MT/s.
- CentOS 7.2 (kernel version 4.10.0).
- 8 Intel® NVMe-based SSDs: PC3700.

The performance comparison tool is the nvme perf tool, located in `example/ nvme/perf` in SPDK package. It can run on a single CPU core with Linux kernel NVMe driver or SPDK NVMe PMD (Figure 16.9).

The nvme perf tool utilizes one core and generates 4KB random data for read I/O on 1, 2, and 8 NVMe SSDs with QD=128. NVMe PMD showcases the performance scaling by driving NVMe SSDs from 1 to 8, but the kernel driver cannot increase the performance while driving more storage devices. The same CPU resource can drive more I/Os with PMD. A single Intel® P3700 800G SSD is about 450K IOPS, NVMe PMD on one CPU core can easily drive the hardware performance limitation of 8 NVMe SSDs (i.e., 3600K IOPS).

From I/O access latency perspective, the total access time can be split as I/O submit and complete. NVMe PMD reduces both I/O submit and complete times, when compared to Linux kernel NVMe driver (Figure 16.10).

Performance evaluation on Intel® Optane SSD

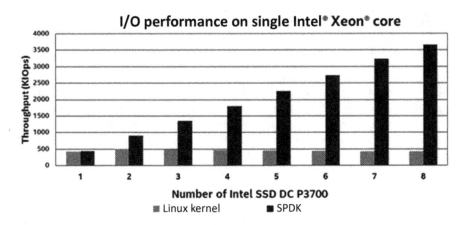

FIGURE 16.9   4 KB random read performance comparison.

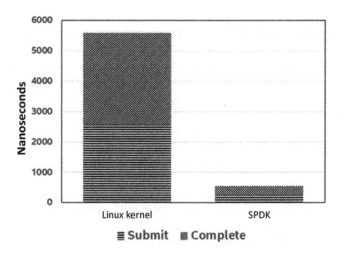

FIGURE 16.10   I/O latency comparison.

Here is a performance test with 3D Xpoint NVMe SSDs on an Intel® Xeon server with two-socket CPU: Xeon E5–2695v4.

- 64 GB DDR4 memory, 8 × 8 GB DDR4 2133 MT/s.
- Ubuntu 16.04.1 (kernel version 4.10.1).
- 1 Intel's NVMe P4800X 375G SSD.

FIO-2.18 is used for performance testing, 4 KB random read with queue depth from 1 to 32, numjobs = 1, direct mode. For testing SPDK, we provide plug-in to FIO; thus, FIO can be directly tested on NVMe PMD. The diagram shows the IOPS throughput and latency at different QDs (queue depths). Only one CPU core is used on the test system. SPDK has higher IOPS at all QDs, and SSD access is saturated at QD = 8. The kernel NVMe driver exhausts all the CPU cycles on a single thread beyond QD = 8. Linux kernel driver needs two cores to reach device saturation (Figure 16.11).

16.3.6   SPDK AND VIRTUALIZATION

While the storage workload runs within the virtual machine (VM), SPDK provides the virtualized framework as shown in Figure16.12. The VM can access the virtio-blk interface to talk with the host (vhost-user-blk in the back-end). This interface is an important abstraction approach to achieving the access without knowing the underlying modules; it is easy for writing the storage application to run the cloud computing infrastructure (Figure 16.12).

16.3.7   DATA SECURITY

Securing "data at rest" is an important step in data security, which is an important feature for a storage system. The stored data are encrypted, so the access need be

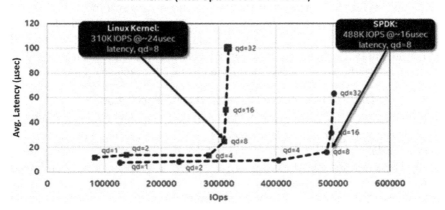

FIGURE 16.11   Performance on Intel® Optane SSD.

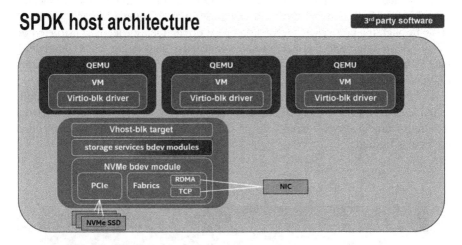

FIGURE 16.12   VM and SPDK acceleration architecture.

authenticated (with credentials such as keys and passwords). CPU (or accelerator) can encrypt the data before they are written to the storage system, or the data are not encrypted before they are written to the storage device (SSD). SSD has the self-encrypting feature which can be enabled to encrypt the data in store. SPDK provides both approaches to securing the data at rest.

16.3.7.1   Cryptodev: CPU/Accelerator-Based Data Encryption
DPDK supports the cryptodev interface, which can support different crypto-imple-mentations (using x86/AES-NI or Intel® QAT acceleration). `ipsec-mb` is a software library that abstracts the x86 AES-NI-based "multi-buffer" crypto-algorithms. As the name suggests, it was initially implemented for IPsec use case. AES algorithms

are applicable for data encryption. A recent academia paper (*Make AES great again*) has claimed that the vectorized AES in the upcoming Icelake server processor will deliver up to 0.16C/B for AES. (It is about 0.64C/B for AES encryption.)

SPDK release 18.01 designed the "crypto vbdev" interface on top of DPDK crypto-dev. This approach is easy to work, because it can always use CPU-based approach. If QAT is present in the chipset (or PCIe card), it can save crypto-processing with QAT accelerator. And the whole access is abstracted by DPDK cryptodev (Figure 16.13).

16.3.7.2 SSD/SED/Opal vbdev: Disk-Based Data Encryption

Opal is an industry standard, driven by TCG (Trust Computing Group). Most SSDs have self-encrypting drives (SED) to encrypt all the data written to the disk and to decrypt the data when the data are read out of the disk. Initially, the disk encryption is only applicable for the whole disk. And the data encryption/decryption is performed by the crypto-unit integrated on the disk. This encryption approach does not occupy the CPU cycles; also, the latest SSD supports the "locking range" to support the multiple data encryptions. Each locking range can have its own key.

The recent SPDK was released in October 2019, and Opal vbdev was introduced. This is in addition to SPDK19.04, and NVMe Opal Library was introduced 6 months ago. The security send/receive request can go through the "Opal Library", which is going through the user space NVMe user space driver. Opal vbdev is constructed on top of NVMe bdev, and the locking range concept is built on top of "Opal vbdev". Intel® Optane P4800 and some Intel® SSDs of P4510 and P4610 are Opal ready (Figure 16.14).

Cryptodev is based on CPU/accelerator. It is flexible to support the user and application level granularity. Opal/SED/SSD is based on disk encryption, and it does not require the precious cycles from CPU or accelerator (more cost), but the encryption use case is limited.

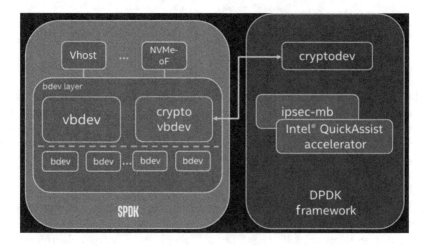

FIGURE 16.13 CPU/acceleration-based data security.

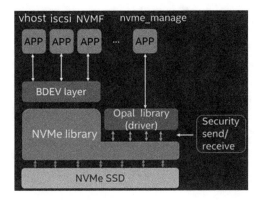

FIGURE 16.14   Opal/SED-based data security.

16.4   CONCLUSION

The hyper-converged infrastructure is a fast-growing business, running on common server platforms, but it offers virtualized computing, storage, and network capabilities. DPDK/SPDK technology can optimize them, in both networking and storage area. DPDK was originally designed to accelerate packet processing and provided a complete set of memory management mechanisms, user-managed CPU scheduling, and user-managed device management mechanisms. The similar practice can also be applied to the storage system to improve the performance of the networks and storage systems.

Index